Ten Questions

Ten Questions
A Sociological Perspective

Second Edition

Joel M. Charon
Moorhead State University
Moorhead, Minnesota

Wadsworth Publishing Company
Belmont, California
A Division of Wadsworth, Inc.

Sociology Editor: *Serina Beauparlant*
Editorial Assistant: *Jason Moore*
Production Editor: *Michelle Filippini*
Text Designer: *Ann Butler*
Print Buyer: *Diana Spence*
Permissions Editor: *Peggy Meehan*
Managing Designer: *Andrew Ogus*
Copy Editor: *Steven Summerlight*
Technical Illustrator: *Susan Breitbard*
Cover Designer: *Andrew Ogus*
Compositor: *Steven Bolinger, Wadsworth Digital Productions*
Printer: *Malloy Lithographing, Inc.*

*This book is printed on
acid-free recycled paper.*

International Thomson Publishing
The trademark ITP is used under license.

Printed in the United States of America
2 3 4 5 6 7 8 9 10—99 98 97 96 95

Library of Congress Cataloging-in-Publication Data
Charon, Joel M.
     Ten questions: a sociological perspective / Joel M. Charon.— 2nd ed.
       p.   cm.
     Includes bibliographical references and index.
     ISBN 0-534-23814-9
     1. Sociology 2. United States—Social conditions. I. Title.
HM51.C457   1995
     301—dc20   94-16174

Contents

Preface

Sociology is a perspective. That is, it is one way of thinking; one way of looking at and investigating the universe. It focuses on the human being as a member of society, so its questions should have importance to all of us who seek an understanding of who we are.

Ten Questions: A Sociological Perspective, Second Edition, is written for students in introductory sociology courses and for students who do not have enough time to take an entire sociology course but who still wish to understand how sociologists think. It is written for sociologists, who sometimes forget the excitement of sociology as they become so involved in the tasks of teaching and research, and for critics of sociology, whose criticisms are too often without foundation. It is for English teachers, physicists, psychologists, artists, poets, and scholars, whose lives are filled with the same questions but whose approaches differ, and for all people who value education and believe, like the Greeks, that "the unexamined life is not worth living."

This book introduces the perspective of sociology by posing ten questions and then answering them, thereby revealing the sociological approach. Sociologists wonder about these questions regularly and most debate them with colleagues, students, or, at the very least, with themselves.

How do sociologists study society?   Can we even see it? Won't our personal bias get in the way? Can we be scientific? What does it mean to be scientific?

What does it mean to be human?   Is human nature something we possess at birth? Is our intelligence the key to humanity? Or is it our language, society, or culture? Just what is the human being, anyway?

How is society possible? What keeps it going? How is it possible for us to cooperate? Is it fear and force that keeps us together, or do we simply like being around others?

Why are people unequal in society? It is human nature? Is inequality built into the nature of society? Is it possible to create a society of equality?

Why do we believe what we do? Are we in control of our beliefs? How does society shape them? Are our beliefs our own?

Are human beings free? Do we control our own lives? What are some of the forces that influence the decisions we make? What is freedom, anyway? Why is it so important for us to think we are free? Are some of us freer than others?

Why can't everyone be just like us? Why do we want others to be just like us? What are values? What is ethnocentrism? Is it good or bad? Are human beings really so different from each other?

Why is there misery in the world? What causes human problems? Why is life so difficult for people? How does society create its own problems? Is it possible to build a better society? Is misery inevitable?

Does the individual really make a difference? Is this simply wishful thinking? When can the individual really make a difference? What works against it? Why does society change?

Is sociology important? Important for what? Will it bring about a better world? A better understanding of self? A better understanding of society? Does it tell us anything important?

These are the ten questions that make up the chapters in this book. They are the most important questions that sociology helps me to answer. They are also the questions that make my intellectual life exciting.

When I revised this second edition I knew from the beginning that I wanted to consider one more question. However, my editor and I did not want to change the title of the book to *Eleven Questions*, nor did we want to take out one of the original ten. We decided to put the eleventh question at the end and to call it an "Afterword." In truth, however, it is the eleventh question in a book entitled *Ten*

Questions. As you will find, it is well-placed at the end of the book, because it can be easily inserted anywhere among the questions. Some will find it best to start with this question, some may use it in the middle, and still others may read it as a last chapter. It is a very difficult but important question. It addresses the most basic question that students have when they first encounter sociology: Because all human beings are unique, is it acceptable to generalize about them? (Isn't generalizing the same as stereotyping?) Shouldn't everyone be treated as an individual? Sociologists, psychologists, anthropologists, and all social scientists generalize about people and society. Are they doomed to failure right from the beginning? The eleventh question is stated as simply:

Should we generalize about people?

The reader will find the answer difficult, interesting, and useful.

A basic assumption underlies this book—that students will enjoy discussing and wondering about such questions. They will recognize education to be more than accumulating facts. And if challenged to debate issues that shed light on the human being, students will discover a fervor in learning that is too often ignored.

This was an exciting project for me. It forced me to make explicit my assumptions about the nature of sociology. The encouragement I received from reviewers was gratifying, and their suggestions for improving the manuscript were invaluable. Especially important were Arnold Arluke, Northeastern University; Laura Fischer-Leighton, Vanderbilt University; Laurel Graham, University of South Florida; Rachel Kahn-Hunt, San Francisco State University; Craig B. Little, SUNY-Cortland; David E. Olday, Moorhead State University; Ronnelle Paulsen, University of Texas, Austin; and Robin Roth, Lesley College.

The Afterword was difficult to write, but several friends were very helpful. I am especially indebted to Ted Gracyk, Mark Fasman, Ralph Levitt, Helen Levitt, Bill Jones, my nephew Michael Segal, and my editor Serina Beauparlant. The following reviewers contributed to the final product of this second edition through their criticisms and recommendations: David Baker, Riverside Community College; Paul Kingston, University of Virginia; Harry LeFever, Spelman

College; William Levin, Bridgewater State College; Donileen Loseke,
Skidmore College; Ginger Macheski, Valdosta State University; Eliz-
abeth Meyer, Pennsylvania College of Technology; Alan Mock, Lake
Land College; Daniel Santoro, University of Pittsburgh, Johnstown;
Kay Snyder, Indiana University of Pennsylvania; Becky Thompson,
Memphis State University.

I would like to thank my editor Serina Beauparlant, whose encouragement and confidence in me contributed a great deal to the
final product.

I owe much to all those people in my life who debated these
questions with me and introduced me to the complexities and wonders of the sociological perspective: my professors at the University
of Minnesota, my fellow students in graduate school, and some of
my colleagues at Moorhead State University. It is wonderful to discuss these questions and create a real commitment to sociology as a
perspective.

I would like to dedicate this book to my wife, Susan, who helps
make my life worth living, and to my sons, Andrew and Daniel,
whose individuality makes me feel proud.

Introduction

All good questions deserve good answers. Good answers require knowledge. Good teachers give good answers, good parents do, and so do good lawyers, tax advisers, doctors, and candidates for office.

A good academic discipline also gives good answers. Often, however, these answers are tentative, qualified, and complex, and they sometimes raise more questions than they answer. That is because almost all academic knowledge results from evidence that is painstakingly gathered and then repeatedly criticized. Answering academic questions entails care, debate, and uncertainty, whether we are dealing with physics and chemistry, art and mathematics, or philosophy, psychology, and sociology.

That is one reason this book was so difficult to write. I did not want to misrepresent the discipline of sociology, for which I have great respect. I realized that many of my answers were far too simple and that I would be hard put to find all sociologists—or even most—fully agreeing with them. Throughout, I worried, Are these the answers that most sociologists would give? By now, however, I realize that *Ten Questions* is *not* a book of answers. *It is much more a book about thinking.* Although all sociologists might not agree with the answers in this book, most would probably agree that it describes how sociologists think.

This book does not describe either the specific ideas that sociologists examine or the many scientific studies that characterize the discipline. It does not present the various specialties and schools of thought that make up the discipline, and it does not show how sociologists disagree on many matters. Instead, it is intended to be an

1

intellectually tantalizing introduction to a way of thinking that you can apply to your most important concerns. The ten questions I consider are among the most important ones that sociologists investigate. Indeed, they are fascinating questions that thinking people will investigate throughout their lives. They form the basis for much of what a serious education should investigate.

As you will shortly see, the sociological perspective is different from your usual way of seeing. We live in a society that emphasizes the individual and tends to look for the reasons for action within individuals. Our religious and political heritage and our tendency to focus on psychology too often cause us to overlook the importance of society in understanding human life. Whereas most people emphasize personality, character, heredity, and individual choice when they discuss human beings, the sociologist keeps crying out to us: "Don't forget society!" "Remember, human beings are social, and that makes a difference in what we all are."

I cannot escape the power of sociology to affect the way I think. Like almost everyone else, I am repulsed by violent crimes. Injustice and inhumanity upset me. War and murder, exploitation and physical abuse, racism and sexism, theft and the destruction of property, feeding other people's addictions and refusing to help the poor—all of these anger me, and frankly, that anger caused me to become a sociologist. But sociologists' approaches to such problems are different from those of most others. Again and again we ask: What kind of *society* does this? What *social conditions* cause individuals to lose their humanity? What are the *social causes* of poverty, crime, and destructive violence?

Whenever I read or hear about a horrible crime, my first reaction is, "What a horrible thing to happen!" My second reaction is, "How can people do that? What's wrong with them?" But then I often get to a third reaction, one that takes more self-discipline and care: "What are the underlying reasons for such acts?" "From what kind of world does inhumanity such as this arise?" As a sociologist I am driven to understand the nature of society (including my own), and I try to appreciate all the different ways in which society affects the human being. Of course, I know that this is not the only way to understand human action, but I believe that it goes a long way.

To introduce the sociological perspective, as mentioned, I have decided to focus on ten questions. To me, questions such as these make sociological investigation exciting. When all is said and done, when we look at all the studies and all the detailed sociological knowledge we have gathered for 150 years, these are still the questions that stand out and that excite discussion and argument within the discipline: How can sociologists study society? What does it mean to be human? How is society possible? Why is there inequality in the world? Why do we believe what we do? Are human beings free? Why can't everyone be just like us? Why is there misery in the world? Does the individual really make a difference? Is sociology important?

Throughout this book I will draw from the works of several important sociologists. These are the writers who have had the greatest impact on my own thinking. Their ideas are the most exciting and meaningful to me, so I will briefly introduce them at this time.

Sociology owes much to the work of Karl Marx (1818–1883). Marx, of course, is best known for *The Communist Manifesto* (1848) and *Das Kapital* (1867), both of which are critiques of capitalism and the rest of society as he knew it. Marx was dissatisfied with how his society functioned, and out of that dissatisfaction (which really amounted to great anger), he developed a theory of society that focuses on social class, social power, and social conflict. Marx's analysis is challenging to what most Americans believe, and he brought to sociology a critical and sophisticated approach to understanding society. Underlying all that he wrote was the idea that *social inequality* is the key to understanding society.

No one has influenced the development of the sociological perspective more than has Max Weber (1864–1920), a German social thinker best known for *The Protestant Ethic and the Spirit of Capitalism* (1905). In this work he shows us that Protestant religious thinking was a central contributing factor to the development of capitalism in the West. Like almost everything else he wrote, this book exhibits Weber's interest in describing the importance of *culture* in influencing how people act as they do. People behave the way they do, he argues, because of this shared belief system, and the only way in which social scientists can understand a people's

actions is to understand their culture. That is why he is so important for the study of religion, modernization, legitimate authority,
bureaucratization, science, and tradition, all particular ways of
thinking that characterize people living together. If we think of
Marx as the critical sociologist, we should think of Weber as the
cultural sociologist. This view is slightly misleading, however, because
Weber was broader than that; he, like Marx, was deeply interested
in social class, social power, and social conflict.

When the name of Emile Durkheim comes up in discussion,
my thinking immediately shifts to "social order." Durkheim (1858–
1917) was driven to understand all the various ways in which society is able to work as a unity. Society, he maintains, is not simply a
bunch of individuals; it constitutes a larger whole, a reality that is
more than the sum of the individuals who make it up. What keeps
it together? How is this unity maintained? Durkheim documents
the important contributions of religion, law, morals, education, ritual, the division of labor, and even crime in maintaining this unity.
Every one of his major works examines it. His most famous work,
Suicide (1897), for example, shows how very low or very high levels
of social solidarity result in high suicide rates. His last important
work, *The Elementary Forms of Religious Life* (1915), documents the
importance of religion, ritual, sacred objects, and other elements of
the sacred world for social solidarity. Durkheim also contributed
greatly to our appreciation of the influence of social forces on the
individual, from suicide to knowledge of right and wrong.

In many ways sociology owes its perspective to the work of
Marx, Weber, and Durkheim. Two other sociologists, both from the
United States, appear now and then in the following chapters. Both
have taught me much about the social nature of the human being,
and especially about the power our social life has over the way we
think. George Herbert Mead (1863–1931), a social psychologist who
taught at the University of Chicago, has been extremely important
in helping me understand the many complex links between society
and the human being. His most important contribution to sociology
is the book *Mind, Self and Society* (1934), which was written from his
lecture notes by devoted students after his death. Throughout his
work certain questions are addressed over and over: What is human

nature; that is, what characterizes the human being as a species in nature? How does society shape the human being? How does the individual, in turn, shape society? Mead persuasively shows that human beings are unique because of the way they use symbols to communicate and to think about their own acts and the acts of others. They are also unique because of their ability to reflect on themselves as objects. To Mead, symbol use, selfhood, and mind are qualities that create a being that can change society and not simply be passively shaped by it. The individual's relationship with society is complex, however, because symbols, self, and mind are socially created qualities, possible only because we are social beings.

The other American sociologist is Peter Berger (b. 1929), who, along with Mead, has had a tremendous impact on my own thinking about the meaning and importance of sociology. *Invitation to Sociology* (1963) and *The Social Construction of Reality* (1966) (written in collaboration with Thomas Luckmann) describe sociology as a special type of consciousness, a perspective that is profound, unusual, critical, and humanistic in its concerns. To Berger, sociology is liberating because it helps to reveal our taken-for-granted realities for what they are: social creations that appear true on the surface but on closer inspection are usually found to be partially true or even untrue. In all of his work Berger shows the power of society to shape human action and thought. Society socializes the human being to accept its ways. To understand the power of society is, for Berger, the first step toward understanding who we are and what we can do to control our own lives.

What I write here is hardly ever my own idea, because it is so heavily influenced by people such as Marx, Weber, Durkheim, Mead, and Berger. This book is inspired by all of these thinkers— and others—so I hope that if you like it, you will turn your attention to their works. I also hope that you will find in them the inspiration that I have found.

REFERENCES

The following works are excellent introductions to sociology, social theory, or specific social thinkers introduced in this chapter.

Berger, Peter 1963 *Invitation to Sociology.* Garden City, NY: Doubleday.

Berger, Peter L., and Thomas Luckmann 1966 *The Social Construction of Reality.* Garden City, NY: Doubleday.

Collins, Randall, and Michael Makowsky 1983 *The Discovery of Society.* 3rd ed. New York: Random House.

Coser, Lewis A. 1977 *Masters of Sociological Thought.* 2nd ed. New York: Basic Books.

Cuzzort, R. P., and E. W. King 1976 *Humanity and Modern Social Thought.* Hinsdale, IL: Dryden Press.

Durkheim, Emile 1893 *The Division of Labor in Society.* 1964 ed. Trans. George Simpson. New York: Free Press.

Durkheim, Emile 1895 *The Rules of the Sociological Method.* 1964 ed. Trans. Sarah A. Solovay and John H. Mueller. New York: Free Press.

Durkheim, Emile 1897 *Suicide.* 1951 ed. Trans. and ed. John A. Spaulding and George Simpson. New York: Free Press.

Durkheim, Emile 1915 *The Elementary Forms of Religious Life.* 1954 ed. Trans. Joseph Swain. New York: Free Press.

Inkeles, Alex 1964 *What Is Sociology? An Introduction to the Discipline and Profession.* Englewood Cliffs, NJ: Prentice-Hall.

Marx, Karl 1845–86 *Selected Writings.* 1956 ed. Ed. T. B. Bottomore. New York: McGraw-Hill.

Marx, Karl 1867 *Capital.* Vol. 1. 1967 ed. New York: International Publishers.

Marx, Karl, and Friedrich Engels 1848 *The Communist Manifesto.* 1955 ed. New York: Appleton-Century-Crofts.

Mead, George Herbert 1934 *Mind, Self and Society.* Chicago: University of Chicago Press.

Ritzer, George 1988 *Sociological Theory.* 2nd ed. New York: Alfred A. Knopf.

Turner, Jonathan H. 1986 *The Structure of Sociological Theory.* Chicago: Dorsey Press.

Weber, Max 1905 *The Protestant Ethic and the Spirit of Capitalism.* 1958 ed. Trans. and ed. Talcott Parsons. New York: Scribner's.

Weber, Max 1924 *The Theory of Social and Economic Organization.* 1964 ed. Ed. A. M. Henderson and Talcott Parsons. New York: Free Press.

How Do Sociologists Study Society?

Researching the Social World

The Beginnings of Rational Proof

We are all indebted to the ancient Greeks (whose civilization reached its peak about 300 B.C.) for the great number of contributions they made to the world. They left us masterpieces in sculpture, pottery, architecture, and drama. They influenced the history of mathematics, science, literature, and democracy. But their special approach to understanding reality is what interests us here. Their culture began and encouraged the study of philosophy. To this day, the contributions of Socrates, Plato, and Aristotle are unparalleled in the history of thought. It is not the ideas alone that stand out, however, but their critical approach, their questioning attitude, that is central to Western thought.

In their search for understanding, the Greek philosophers came to reject authority, or what culture taught, or what people believed in their heart. Instead, they developed a measuring stick to judge ideas, what we now call *rational proof*. To understand the significance of rational proof, think of a simple measuring stick. We determine the length of a line by putting the measuring stick (divided into inches or centimeters) next to it. The stick is a standard used not only to measure that line but also to compare it with others. Based on our measurements, we can rightly say, "This line is longer than that line."

The Greeks developed a "measuring stick" for determining the truth or falsehood of an idea. They created standards of proof to evaluate the quality of reasoning that had gone into a conclusion to determine if that conclusion was correct or in error. Before the Greeks, the truth of an idea was determined by who said it, where it was written, or simply what the idea was. In many groups, truth or falsehood is still determined by such standards. For example, the leader of a country is likely to believe something if the idea is patriotic or if it is told to him or her by an admirer. In some religious or political groups something is thought to be true if the leader says so or if it conforms to the group's written sources. It is tempting for all of us to accept something as truth that is clearly expressed, comfortable, or presented by someone we like.

The Greek philosophers were wise, however. The truth or falsehood of an idea, they taught, has nothing to do with what is already believed or with what individuals or groups declare to be right. It is not determined by gut feelings or intuition, majority vote, or the wise people in a society. Simply put, the Greek philosophers taught us that ideas are correct if we arrive at them through a careful reasoning process, if we *prove* them through the rules of logic. Honesty is at the heart of sound proof: do not twist your thinking to prove what you want to prove; do not exaggerate, jump to unwarranted conclusions, or scream that an idea is wrong simply because it disagrees with what you already believe. Instead, show us through a careful, organized, and honest examination of the idea. Take it apart: look at its assumptions, search for its contradictions, examine it according to what else has been proved, dissect it into its components, and examine those components. If an idea can be supported through such activities, we can be relatively sure that we have arrived at something approximating the truth. If not, it should be rejected, no matter who says it, no matter how much we want to believe it, and no matter whether it is what we have always been taught.

Greek civilization eventually declined, and its influence in the Western world was gradually replaced by an all-powerful Christian church that became the source of all truth. From around A.D. 300 until 1400, the most important approach to truth was a spirit of faith. People were expected to accept what their church taught rather than to seek wisdom on their own. The spirit of critical, rational analysis of established ideas was defined as heresy and punished.

Eventually, however, the spirit of Greek philosophy became acceptable again in the intellectual community. With the rise of a critical philosophy and science around the fifteenth century, it became a powerful force in the Western world. The development of social science in the eighteenth century and the founding of sociology as a social science in the nineteenth century were consequences of these developments and became part of this critical tradition.

Proof, Science, and Sociology

The Need for Scientific Sociology

Whenever I try to explain sociology to a mathematician friend, he gives up and declares, "I think higher math is easy compared with understanding human beings." When we puzzle over personal, social, and worldly problems, it becomes obvious that human beings are difficult to make generalizations about simply because of their complexity. But this complexity is not the only thing that makes sociology such a difficult discipline. People walk into sociology, unlike mathematics, already believing that they know a great deal about the subject at hand: human beings and society. What could be more familiar? It is difficult to get people to question what they already "know"; it is more difficult still to get them to recognize that what they know is not necessarily true.

Think about it for a moment: where have our ideas about human beings and the nature of society come from? We are embedded in a society that teaches us to accept its fundamental ideas and values, and those who teach us have a stake in that society and its belief system. In short, we are all cultural beings. We learn the basic ideas, values, and norms of our society and incorporate them into our images of ourselves and our social world. The beliefs of a culture need not be true or proved; we normally accept them without serious question because they are what we inherit from our society's past. Socialization in our early years makes us into functioning members of society. By virtue of this primary socialization we internalize the ideas of those immediately around us. As young children we know no better: we must depend on our parents and others close to us to teach us what they know. As we get older, school, television, friends, and other agents of socialization introduce more ideas, and many of these ideas reinforce the earlier ones. When we walk into a university lecture class and encounter a professor who questions those ideas, or when we come across a person with a different point of view, we may be influenced to change our ideas, but normally the burden of proof is on the new, not the old. We demand convincing "evidence" before we are willing to reject the ideas that come from our primary socialization. *Most people find it difficult to seriously evaluate their ideas about the human being and society through applying the measuring stick of the Greek philosophers because those ideas have become so much a part of them through socialization.*

This questioning, however, is exactly the purpose of sociology. Auguste Comte (1798–1857), the French thinker who coined the term *sociology*, argues that the critical methods of the ancient Greeks can and should be applied to society. In fact, these methods should go even further: they should rely on a strictly scientific approach, a measuring stick even more demanding than rational proof. He believed that he was founding an important discipline whose purpose would be to analyze carefully and objectively the nature of society, to question, if need be, what culture teaches us about society, and to arrive at ideas supported by evidence rather than by what certain people believe or by what has been handed down to us from the past. From Comte to the present, this has been sociology's central goal, and its basic purpose has always been to look beyond what people have been taught about society to what actually exists (irrespective of what we might want to exist). This is the reason for sociology, and this is its strength.

Empirical Proof

The heart of rational proof is the recognition that the basis for truth must be found in reason, in a careful appraisal of ideas. Socrates, one of the greatest of Greek philosophers, best represented this recognition. The Socratic method of investigation is a continuous set of questions posed to someone. Through this method he revealed to the individual the assumptions that people made without careful thought, the illogical conclusions that they reached, and the poor evidence that they relied on for their set of beliefs. Socrates, through his questioning, did not discover truth so much as he uncovered untruth, and in so doing he caused others to seek truth in a more careful, thoughtful way.

Actually, this emphasis on proof laid the foundation for modern science. Philosophical questioning and the idea that conclusions must be questioned and must meet the measuring stick of rational thinking developed into what came to be known as *empirical proof.* (Empirical proof uses *careful observation,* rather than careful thought, as the basis for measuring the truth or falsehood of an idea.) In fact, some people trace empirical proof to Archimedes, a renowned Greek thinker and one of history's first scientists. Archimedes wanted to know how he could measure the volume of a mass (such as a king's crown). After all, one could not measure the mass of a crown with a measuring stick, since it was such an impossible shape. One day, while taking a bath, he noticed that the water rose in the tub when he got in and lowered when he got out. That, in a flash of insight, was the answer to his problem: to measure the volume of anything, all one must do is measure how much water is displaced by the object. The story is told that Archimedes ran naked into the streets crying out, "Eureka, Eureka, I have found it!"

How did Archimedes find his answer? Simply put, *he observed it.* This example illustrates well what empirical proof is: it is proof that is *observed.* This type of proof eventually became the basis for all the sciences, from physics and biology to psychology and sociology.

Rational proof and empirical proof are both measuring sticks. It is not what people believe that determines truth; it is *how they arrive at their beliefs.* Can it be *proved,* rationally or empirically, or is it accepted simply because someone says so or because it is written down? To prove something rationally or empirically means that one is forced to clearly describe in a step-by-step process how one arrived at one's conclusions so that others can check out the proof— so that they can analyze the evidence, bring forth new evidence, or simply repeat the process to see how good the evidence actually is. This is the basis for understanding in both philosophy and science. In a sense, this is a democratic process, because proof is not owned by anyone. Whatever is known must be put on the table and shared with others so that they, too, have a chance to measure it.

Empirical proof is the basis for sociology and for many of the ideas in this book. It is the basis for the conclusions in the various specialties of sociology, from the study of family and religion to the study of revolution and culture. Observation is the basis for science, and so it is the basis for sociology. Acceptable evidence is that which can be *observed* by one individual and *shared* with others, so that they can observe it, check it, criticize it, build on it, or disprove it. Observation can take place in a laboratory or in the natural environment. We can observe prominent people, items checked on a questionnaire, or diaries and letters. We can observe people in a gang, a corporation, a religious group, a football team, or an army. We can observe the speeches of political leaders, articles written in newspapers, textbooks, or magazines. Sometimes observation is a relatively easy matter—for example, seeing how many people marked letter *A* for item 7 on a questionnaire or how many people under 21 committed suicide in Minnesota during 1993. Sometimes observation is far more difficult, and researchers must be especially imaginative and careful in their observation. How do poor people go about finding work? How do wealthy people exert power in government? What things do little boys say to little girls that reflect their understanding of gender roles? No matter what the question, if sociology is being done, researchers must use empirical evidence to support their ideas, evidence that can somehow be observed and therefore can be checked by others.

Two examples of empirical work in sociology will give you some idea of how sociologists "observe." Durkheim's study of the causes of suicide is the first example. He was concerned about the high suicide rates in many communities and societies in Europe at the end of the nineteenth century. He could "see" those rates by simply looking at the number of suicides per 100,000 people in a given population. He found a remarkable consistency in the rate of suicide in France over time, and he found that in comparison with other societies, France's rate was higher than some and lower than others. He wanted to know why rates differed among societies and why they differed among communities within a society. He theorized that these rates were heavily influenced by the level of social solidarity in the community—that is, by how integrated the community was. He could not "see" the level of solidarity, so he applied what he logically thought out about solidarity in various communities (and he expresses exactly what he is thinking to the reader of the study). He argues, for example, that Catholic communities will have higher levels of social solidarity than Protestant communities, because Catholics are more embedded in a church whereas Protestants emphasize the individual's relationship to God. He further argues that the Jewish community is more integrated than either the Catholic or Protestant one (because Jews in the Europe of 1900 were more separated from the larger community and their religion permeated every aspect of their daily living).

Then Durkheim was ready to "see" the evidence concerning suicide rates. He found what he expected: Protestant communities had the highest suicide rates, and Jewish communities had the lowest. He contrasted other communities: urban versus rural and college-educated versus non-college-educated, for example. In every case the more individualistic communities had higher suicide rates. Durkheim observed data collected by the government—imperfect, incomplete, and perhaps even biased data. Of course, one need not believe his theory about social solidarity and suicide rates. But the beauty of science is that one knows how he thought and how he observed. One can go and observe the same data to check him out, or one can show that the same data can be understood in another way. One can now examine data in the United States or in any other part of the world. One does not have to take Durkheim's word for anything.

The second example is a study published in 1977 by Rosabeth Kanter. Kanter was interested in how the large organizational system, by its very nature, worked against equality between men and women. She knew that there were many ways to "observe" men and women in the corporation: she could take a nationwide survey, she could interview presidents of corporations, or she could examine existing data on how many women were in managerial positions or on boards of directors. What Kanter decided to do, however, was an in-depth study of one corporation. She sent out a mail survey to a sample of sales workers and sales managers. She interviewed many employees about their work and their positions in the corporation. She systematically examined 100 appraisal forms filled out on secretarial performance. She attended group discussions, observed training programs, and examined many documents within the corporation. She informally visited with employees at lunch, in hallways, and wherever else she could meet them. The success of the study, of course, depended on how well she could convince other social scientists that her methods had been careful, objective, and thorough. The strengths of her study were the depth that she was able to achieve in one corporation and the diverse ways in which she observed. Of course, the weakness was that she had enough time and money to study only one corporation. We need not believe what she found, however: We can do another case study. We can compare what she found in her observations with what others found in theirs. We can check out national surveys or do our own national survey, or we can examine already collected government data on all corporations.

These are not necessarily the best studies in sociology, but they are good examples of what is meant by *empirical evidence*. W. I. Thomas studied the Polish peasants who came to the United States through examining diaries and letters. Frederic Thrasher and his associates studied more than a thousand gangs in Chicago through observing their actions on the street and interviewing members. W. H. Sewell wanted to understand how social class affected success in the American system of education, so he followed students through the system by giving them questionnaires over many years. Gary Fine studied preadolescent socialization through observing how boys acted on Little League baseball teams.

Observation in Sociology

Sociology is a particular type of science. It is not easy to observe groups, societies, power, interaction, or social class, because these are not physical entities like leaves, skin, rock, or stars. Nor is it easy to observe people's ideas or values, their morals or their hopes. Thus, the scientist must watch how human beings present themselves to others—what they do, what they say, and what they write—and must then look beyond and infer the existence of a more abstract social reality. Thus, when people act together, we infer the existence of a group and then draw from our observations the qualities of groups, the ways in which they form and function and their effects on individual action. When Durkheim tried to understand society, for example, he focused on people's rituals, the moral outrage they exhibited toward certain individuals, and the objects they worshiped. He showed how these acts and beliefs revealed the power of society over human action, and he showed their necessity for the continuation of society.

Sociologists do not have a narrow, rigid view of science. On the contrary, their view is that science must be open and that its techniques must be varied. They recognize that certainty is almost impossible and that their ideas must thus remain tentative. Max Weber contends that all scientists must be prepared to see their own ideas overturned with new evidence in their own lifetimes, especially scientists who study the human being. He emphasizes that there are exceptions to almost every conclusion we make but that they do not negate the conclusion. They do make it more tentative and complex. For example, people generally end up in the social class in which they are born—but not everyone does. We must ask why birth is so important to class placement and why there are exceptions. With each conclusion there are new questions and new directions for investigation. And with each conclusion there are some who are skeptical and decide to test it in a slightly different way. It is not final truth that characterizes sociology (and most other science); it is a constant debate in which scientists, through published writings, put forth their ideas and evidence and wait for others to agree or disagree.

The refusal of sociologists to take a narrow view of science is also seen in their diverse and creative research. Sociologists will not get very far by setting up laboratory experiments, using microscopes, running rats through mazes, or mixing chemicals in test tubes. They use laboratory experiments when they can, but such experiments are less common in sociology than in certain other sciences. Although it is desirable to have complete control over your environment in order to have faith in your results, this is usually impossible. We must observe situations over which we lack full control, taking as much care as possible to record what is actually taking place and recognizing that we must be more humble about what we find out than the physicist or chemist. If we want to test an idea about why social change occurs, we can carefully observe events and gather data from newspapers and public speeches in Eastern Europe today and contrast them with similar events, newspapers, and speeches in mid-nineteenth-century Eastern Europe. Systematic observation of people is difficult, and we must develop creative techniques that respect the complexity of the subject—human society—that is being investigated. Thus, human beings are often motivated by ideas, values, attitudes, and moral concerns that cannot be observed but can only be understood through questionnaires, interviews, and analyses of their speech and writing. To understand conflict, cooperation, inequality, agreement, and power, sociologists observe people in groups wherever they can be found, they enter into and study a community for one or two years, or they feed data from newspapers and magazines into computers. They study crime rates, suicide rates, divorce rates, and unemployment rates to understand what qualities characterize a certain society. And they observe everyday rituals and more formal religious rituals to understand how people think about the universe and their own lives. Every single act of the human being is something for sociologists to study, because it can help us make sense out of a larger picture that is not easily seen. We take whatever seems reliable from our observations, preserving data that we have confidence in and discarding ideas that the evidence does not seem to support.

Objectivity in Science

Science is not merely observation; it is *careful observation*. The purpose of science is to exercise control over our observations, to help us determine that what we say we have observed actually "is there" and is not just what we want to see. Weber describes science—and sociology as a science—as "value-free" investigation: an attempt to carefully and objectively observe the world "as it is" rather than as we would like it to be. He means that our only commitment must be to scientific investigation itself; our conclusions always remain open to further investigation.

To be objective means literally to see the world as an "object" apart from ourselves, to separate it as much as possible from our subjective perception. This is more easily said than done, however. Our eyes play tricks on us; we usually see what we want to see. We tend to see things according to what we have learned in our society, according to ideas and values that we are used to and that make us feel secure. Science recognizes this bias and pushes the investigator to overcome it. That is the reason for our reliance on the scientific method, which is nothing more than a systematic and disciplined control over our investigation. It is the reason for the many rules that scientists agree to follow in posing a question, setting up a hypothesis, testing that hypothesis, arriving at a conclusion, and relating that conclusion to the original question posed. Strict rules tell scientists how to create good theory, how to sample, how to accurately observe, how to control the study so that it focuses only on what they want to study, how to carefully interpret data, and how to refine theory on the basis of the evidence. Strict guidelines tell scientists how to report to other researchers the way in which an idea was formed, how a test was developed, what was observed, and how the results were interpreted. All of the rules that are there for the scientist to follow have one basic purpose: to ensure, as much as possible, that the work done is *objective* and that the personal bias of the scientist is minimized.

Of course, total objectivity is impossible. Instead, objectivity is a goal toward which scientists must strive. Total objectivity is probably more difficult to achieve in social science than in natural science and is probably more elusive in sociology than in any other social science. This is because we are all embedded in a society. Representatives of that society have taught us its culture; that culture surrounds us all, telling us what is true and false. We have all learned a bias that we are not fully aware of. To understand the world objectively is to somehow overcome that culture, recognizing it for what it is and seeing through it as much as possible. All science has had to do this. For example, Western culture for a long time taught that the earth was the center of the universe, that nature was static, and that the earth was flat and stood still. Scientists were persecuted and even put to death for showing that these ideas were untrue.

It is even more difficult for sociology to overcome what culture teaches. Our culture supplies us with ideas that we take for granted when we encounter other people and other cultures: that poverty and crime result from lazy or evil people, that our society is a place where everyone can become anything he or she wants, that human action results simply from individual choice and is relatively free of social influence, and that biological differences in race and gender are important to consider when we try to understand human differences. Whatever the truth or falsehood of such ideas, we are steeped in them, which makes it difficult for us to be objective. *This is exactly why a science of sociology is necessary.*

Two Assumptions of Science

Religions make several assumptions about the universe. Most assume that a God exists and that this God has given humankind a set of moral laws to live by. Most of them also assume that people's souls live after death, and they assume that a body of truth has been given to humankind by God.

Science, too, makes assumptions about the universe. The first is that nature is lawful. The second is that natural events are caused by other natural events.

Natural Law in Science and Sociology

To believe that nature is lawful is to hold that nature is governed by predictable regularities. Scientists believe that it is possible to explain the past and, on that basis, to predict the future. It is regularity that governs nature rather than haphazard, unpredictable chaos. We can generalize about events in nature rather than simply believing that each event is unique. The purpose of science is to understand natural law. Scientists are driven to solve the puzzles that are assumed to exist in nature. Before science, people explained nature by the acts of supernatural forces, God or gods intervening and determining who would live and who would die, which wars would be fought and who would win, what progress would be made, and what losses would be suffered. It may be that the universe is controlled by supernatural forces and that natural law is not important, but science—if it is to understand anything—must proceed on the assumption that events occur because of regularities in nature. Why does something happen? The drive of the scientist is to seek answers in natural law.

Sociology is a science, and thus, like other sciences, it assumes natural law. Human beings are part of nature, and they are subject to regularities that can be isolated, understood, and predicted. When people interact, for example, they almost always develop a system of inequality, a set of expectations for each actor (called *roles*), and a shared view of reality (called *culture*). When a society industrializes, there is a strong trend away from tradition and toward individualism, and because the individual is less firmly embedded in social groups, a higher suicide rate results. When an oppressed group's expectations race ahead of what the dominant group in society is able or willing to deliver, there will be violent rebellion, even widespread revolution. Evidence seems to show that such patterns have usually occurred in the past and that they will occur in the future. And we can point to evidence to help explain what other conditions aid their occurrence and why exceptions or different patterns arise.

Thousands of social patterns exist. Sometimes we can discover patterns that are true for everyone; more often we uncover patterns that are true for large numbers of people. We can uncover tendencies, directions, and probabilities in the social world, but it is very difficult to uncover certainties and absolute laws. However, this is also true for other sciences. Not everyone who is exposed to the HIV virus develops AIDS. We cannot predict how long it will take every individual to recover from an illness or die from it. We cannot predict exactly when an earthquake will occur or if a given individual will graduate from high school or college. Economists cannot tell exactly when a depression, deflation, or inflation will occur, and they cannot tell us with much certainty what will occur to the economy if we raise taxes, lower them, or abolish them. However, science has broadened our understanding of all such matters. The more we discover about natural and social patterns that exist in the universe, the more we will gain useful knowledge, and the more we will be able to predict and even shape the future. It is not difficult to see how far we have gone in our knowledge if we simply compare what we know now with what people used to believe about poverty, social change, suicide, alcoholism, racial inequality, gender inequality, crime, and social class. Today, almost all of us see that such situations are caused, at least to some extent, by social conditions, but before the eighteenth century the influence of society on human behavior went unrecognized. Human nature, character deficiencies, or supernatural forces were the only explanations.

Natural Cause in Science and Sociology

The second assumption of science is that events in nature are *caused* by other natural events. Objects fall to the ground because of a natural force that we call *gravity*, which pulls objects toward the center of the earth sphere and which results from the way the earth turns. Germs cause some diseases. Biological inheritance causes certain forms of cancer. Poverty causes some crime. Exploitation of disadvantaged groups helps to create and perpetuate their inferior position in society. The whole purpose of experiments in science is to try to link independent variables (influences, causes) to dependent variables (results, effects) in order to show that when variable X occurs, it will produce variable Y.

   Cause is not easy to establish. Scientists must go through a long and difficult process to uncover it. Note how difficult it has been to establish that smoking is a cause of lung cancer. One must show that smokers are at a higher risk than nonsmokers. One must show that it is not smoking combined with air pollution or eating red meat that causes cancer, but smoking alone. One must show that the more smoking one does, the more likely one is to develop lung cancer. One must show that it is not personality, gender, class, or place of residence that causes both smoking and lung cancer but rather that smoking itself is directly linked to lung cancer. And one must try to go further: What exactly is it about smoking that causes cancer? If one stops smoking, does that reduce the risk? Do other activities or human characteristics increase the risk for smokers?

Sociology applies this principle of cause to the human being. Human belief and action have causes. Other social sciences share this assumption with sociology. Psychology shows us how environment and heredity interact to shape the person and how the qualities of the person, in turn, shape what he or she thinks or does. Economics isolates economic forces in society, political science looks at political forces, and anthropology examines biological and cultural forces. Social psychology shows us how other people around us cause what we do.

Because of the complexity of the human being, this underlying assumption of cause is more difficult to apply with certainty in social science than it is in natural science (i.e., the physical and biological sciences). It is rare that we are able to isolate clear, inevitable causes for what people think and do. We tend to uncover tendencies and probabilities. We are more likely to call a cause in social science an "influence" or a "contributing factor." Such information is highly valuable, as imperfect as it is. Being abused as a child is an important influence on whether one abuses children as a parent. The size of an organization is an important contributing factor to its developing a bureaucratic structure. Being born into poverty affects one's chances to become rich, to be successful in school, and to become the president of the United States.

Sociology has come very far since the nineteenth century toward convincing the general population that social circumstances make an important difference for individual action. Consider, for example, the total absence of a social explanation for suicide before 1900. The work of Durkheim was an important breakthrough: societies have different suicide rates, and those rates are a result of *social forces,* such as the degree of social solidarity or the degree of social change. Most people today recognize that suicide is not simply an isolated decision, that society in many ways influences that decision. Individual acts of crime, individual acts of divorce, and individual choices to have children are partly a result of the social world one lives in. Segregated schools create conditions for unequal education, and sex discrimination in hiring, paying, and promoting creates a segregated and unequal labor force. To understand human life, many people have been influenced to examine social forces.

Sociology: Understanding the Puzzle of Society

Studying sociology is like putting together a big puzzle or solving a great mystery. We gather small parts of the puzzle—for example, the rising divorce rate, the increasing numbers of women in the work force, the greater commitment to individual choice, longer lives—and the parts begin to make sense when we fit them together. There is excitement every time another part fits. We take isolated and seemingly meaningless events (people standing in line at a movie theater; a crowd getting excited at a football game; a revolution in Eastern Europe) and observe and record them. Taken alone, none of the events gives us a key to our mystery, but together the events tell us something important about the nature of crowds, about how crowds sometimes become organized, and about why crowds sometimes become violent. Understanding society is a continuous puzzle, or mystery: it will never be solved, but we can slowly fit together a few parts and gain some new understanding.

It is difficult to introduce you to the complexity of this puzzle—to show you the continuing process of discovery that characterizes sociology. It is tempting to show you the parts without trying to fit them together. After all, the parts are easier to study scientifically. For most sociologists, however, sociology represents far more than the parts—it remains an exciting way of understanding the whole puzzle of society. The purpose of this book is to try to show some of the ways in which sociologists think as they try to understand the puzzle.

Summary and Conclusion

Sociology, by its very nature, is a questioning perspective, a "critical point of view." All science must be suspicious of what people know from their everyday experience, but sociology, especially, must be suspicious, and this suspicion leads to a questioning, probing, doubting, analytical approach to understanding society and the human being. It questions what many people take for granted.

This chapter began with a description of the ancient Greeks and their many contributions to humankind. Probably the most famous

Greek of all was Socrates, a philosopher who was never satisfied with the answers people gave him. Socrates questioned what people thought, forcing them to be critical: "What is goodness?" "What is virtue?" "What is the good society?" The replies people gave were stock answers, which had been learned but rarely thought out. No matter what answer people gave, Socrates had another question that caused further thought. To him, this is what education must be: a continual search for understanding through asking questions and exposing superficial answers, causing the student to grasp an idea through careful examination rather than simply recite what was taught.

This is also the mission of sociologists: to probe the answers people give, to uncover what they believe, to examine reality through controlling personal and social bias, and to see the human being in society as clearly and carefully as they possibly can. Sociologists study the assumptions and problems of capitalism when many people today are claiming it to be humankind's salvation. We study the causes of crime and the problems associated with increasing the prison population when many care only about the evils of crime and the necessity for isolating dangerous individuals from legitimate society. We study the functions of religion in society when many people see religion as part of the sacred world, not to be studied but to be accepted and used for guidance. We study the world of the violent youth gang to understand culture, stigma, and adolescent reactions to society when many people simply want such groups to disappear. We see dimensions of inequality that others ignore, we examine the meaning and contributions of deviance to society as well as its causes, and we try to uncover the purpose of social ritual rather than simply to perform it. Sociology goes beyond the obvious; it asks questions where most people do not; it recognizes (as Peter Berger claims) that the first wisdom is that things are never what they seem.

Putting together the puzzle of society involves a critical approach to understanding, and that is the essence of science. Throughout this chapter I have attempted to show that it is important to study society scientifically because science encourages care and objectivity. This chapter has emphasized three points concerning how sociologists try to understand society:

1. Ideas must be supported by empirical research. Such research must be careful, creative, and diverse.

2. As a science, sociology must constantly attempt to be objective. It must critically investigate ideas that most people have come to accept as part of their culture. This makes sociology a very difficult science.

3. The human being and human society are part of nature, and thus they are governed by a set of natural laws. Human events, it is assumed, are caused by identifiably natural—that is, social—causes.

It is not safe to have sociology around if one wants people to believe in an established system of belief. That is why a critical sociology did not thrive in Eastern Europe and the Soviet Union. That is also why the United States of the '80s and the '90s—bent on converting the world to its ways and encouraging its citizens to care about material progress—has not always been kind to the critical ways of sociology.

If it is important to understand our life honestly, we must look at ourselves; we must examine our ways; we must question what we know and see. This is what a liberal-arts education should be. And this, in the end, is the purpose of sociology as a science.

REFERENCES

The following works contain introductions to science, issues related to sociology as a science, or examples of good empirical studies in sociology.

Babbie, Earl M. 1989 *The Practice of Social Research.* 5th ed. Belmont, CA: Wadsworth.

Becker, Howard S. 1976 *Boys in White: Student Culture in Medical School.* Rev. ed. Chicago: University of Chicago Press.

Bell, Colin, and S. Encel (eds.) 1978 *Inside the Whale: Ten Personal Accounts of Social Research.* Rushcutters Bay, Australia: Pergamon Press.

Berger, Peter 1963 *Invitation to Sociology.* New York: Doubleday.

Blalock, Hubert M., Jr. 1970 *An Introduction to Social Research.* Englewood Cliffs, NJ: Prentice-Hall.

Cohen, Morris R., and Ernest Nagel 1934 *An Introduction to Logic and Scientific Method.* New York: Harcourt Brace Jovanovich.

Durkheim, Emile 1895 *The Rules of the Sociological Method.* 1964 ed. Trans. Sarah A. Solovay and John H. Mueller. New York: Free Press.

Durkheim, Emile 1897 *Suicide.* 1951 ed. Trans. and ed. John A. Spaulding and George Simpson. New York: Free Press.

Durkheim, Emile 1915 *The Elementary Forms of Religious Life.* 1954 ed. Trans. Joseph Swain. New York: Free Press.

Erikson, Kai T. 1976 *Everything in Its Path.* New York: Simon and Schuster.

Fine, Gary Alan 1987 *With the Boys: Little League Baseball and Pre-adolescent Culture.* Chicago: University of Chicago Press.

Glazer, Myron 1972 *The Research Adventure.* New York: Random House.

Golden, Patricia M. (ed.) 1976 *The Research Experience.* Itasca, IL: Peacock.

Kanter, Rosabeth 1977 *Men and Women of the Corporation.* New York: Basic Books.

Kennedy, Robert E., Jr. 1989 *Life Choices.* 2nd ed. New York: Holt, Rinehart and Winston.

Kerlinger, Fred N. 1986 *Foundations of Behavioral Research.* 3rd ed. New York: Holt, Rinehart and Winston.

Kuhn, Thomas S. 1962 *The Structure of Scientific Revolutions.* Chicago: University of Chicago Press.

Liebow, Elliot 1967 *Tally's Corner.* Boston: Little, Brown.

Lofland, John 1966 *Doomsday Cult.* Englewood Cliffs, NJ: Prentice-Hall.

Madge, John 1962 *Origins of Scientific Sociology.* New York: Free Press.

Madge, John 1965 *The Tools of Social Science.* Garden City, NY: Doubleday.

Mead, George Herbert 1934 *Mind, Self and Society.* Chicago: University of Chicago Press.

Mills, C. Wright 1956 *The Power Elite.* New York: Oxford University Press.

Mills, C. Wright 1959 *The Sociological Imagination.* New York: Oxford University Press.

Myrdal, Gunnar 1944 *An American Dilemma.* New York: Harper and Row.

Myrdal, Gunnar 1969 *Objectivity in Social Research.* New York: Pantheon.

Reynolds, Paul D. 1982 *Ethics and Social Science Research.* Englewood Cliffs, NJ: Prentice-Hall.

Riley, Gresham (ed.) 1974 *Values, Objectivity, and the Social Sciences.* Reading, MA: Addison-Wesley.

Sanders, William B. (ed.) 1974 *The Sociologist as Detective: An Introduction to Research Methods.* New York: Praeger.

Sewell, W. H., R. M. Hauser, and D. L. Featherman 1976 *Schooling and Achievement in American Society.* New York: Academic Press.

Simon, Julian L. 1985 *Basic Research Methods in Social Sciences: The Art of Empirical Investigation.* 3rd ed. New York: Random House.

Smith, H. W. 1981 *Strategies of Social Research: The Methodological Imagination.* 2nd ed. Englewood Cliffs, NJ: Prentice-Hall.

Stinchcombe, Arthur L. 1968 *Constructing Social Theories.* New York: Harcourt Brace Jovanovich.

Thomas, William I., and Florian Znaniecki 1918 *The Polish Peasant in Europe and America.* 1958 ed. New York: Dover.

Thrasher, Frederic 1927 *The Gang.* Chicago: University of Chicago Press.

Wallace, Walter L. 1971 *The Logic of Science in Sociology.* Chicago: Aldine-Atherton.

Wallace, Walter L. 1983 *Principles of Scientific Sociology.* Chicago: Aldine-Atherton.

Weber, Max 1905 *The Protestant Ethic and the Spirit of Capitalism.* 1958 ed. Trans. and ed. Talcott Parsons. New York: Scribner's.

Weber, Max 1919 "Science as a Vocation." In *Max Weber: Essays In Sociology.* 1969 ed. Trans. and ed. H. H. Gerth and C. Wright Mills. New York: Oxford University Press.

Whyte, William Foote 1955 *Street Corner Society.* Chicago: University of Chicago Press.

Zimbardo, Philip 1972 "Pathology of Imprisonment." *Society* 9:4–8.

2

What Does It Mean to Be Human?

Human Nature, Society, and Culture

*T*he Twilight Zone" was an exciting and sometimes eerie television series that was popular in the 1960s. As an audience we seemed to know that a surprise—scary, wondrous, or both—awaited us at the end if we patiently followed the story. One of the episodes that has stayed with me concerned a journey by American astronauts who landed on a distant planet. They befriended the inhabitants (who looked very human) and were pleased to find themselves in a luxurious home, much like one they might have had on earth. However, they eventually became aware that they could not leave the home, that they had become prisoners. Then a wall opened up and revealed a large pane of glass with spectators peering in. The astronauts were on display under the label "*Homo Sapiens* from the Planet Earth."

Since then, I have been bothered by a question that probably few people asked after seeing that episode: What would those creatures from earth that we call human beings have to do in the cage for those outside to understand what human beings are really like? Phrasing the question differently: What is the human being? What makes us "human" and not something else? In what ways are we like all other living creatures? What do we have in common with other animals? How are we different? Of course, these questions have probably teased the thinking person from the very beginning of human existence. Look around. We see worms, dogs, cats, bees, ants, and maybe fish. Are we unique? All species of animals are unique. But how are we unique? What is our essence as a species? What would the astronauts in the cage have to do to reveal the essence of the species they represent?

We might begin by recognizing that we share many qualities with other animals. Human beings are mammals, which means we are warm-blooded, we give birth to live young, the female nurses the young, and we have hair covering parts of our body. We are also primates; therefore, we are mammals who are part of an order within nature that is characterized by increasing manual dexterity, intelligence, and the probability of some social organization.

Philosophers have made various claims about what our outstanding characteristic, our key quality, is. They have pointed to our ability to make and use tools, to love, to know right from wrong, to feel, to think, or to use language. Religious leaders emphasize that we have a soul and a conscience. They may also stress that we are created in God's image (thus, we are closest to God) or that we are selfish and sinful (thus, we are similar to other animals). The more cynical critic maintains that we are the only animal that makes war on its own kind (even though other animals are clearly aggressive toward members of their own species).

Psychologists may focus on the fact that humans are instinctive, that they are driven by their nonconscious personality, that they are conditioned like many other animals, or that, unlike other animals, they act in the world according to the ideas and perceptions they learn. Most will maintain that human beings develop traits early in life out of an interplay of heredity and environment.

Sociologists, too, have much to say about the nature of the human being. They maintain that our unique qualities are that we are

1. *social*, in that our lives are linked to others and to society in many complex ways;

2. *cultural*, in that what we become is not a result of instinct but of the ideas, values, and rules developed in our society.

Without these two core qualities, we would not be *what we are*. Put us in a zoo, take away either of these qualities, and visitors to the zoo would see something very different. To understand human beings as a species, therefore, it is important to understand how these two core qualities enter into our lives. It is also important to recognize the complex interrelationship between the social and the cultural: our culture *arises from* our social life, and the continuation of our social life *depends on* our culture.

Human Beings Are Social Beings

Many animals are social beings in a general sense. Fish are social in that they swim in schools, probably for protection. Bees and ants are better organized than any human society. Our closest relatives, apes and monkeys, are social, and their social life is similar to ours.

To claim that the first human beings were social is simply to recognize that our social life was always important to us and that humans never existed without this quality. The first humans were not isolated individuals but beings who interacted, were socialized, depended on one another, and lived their whole lives around others. Of course, some may have chosen an existence apart from others as they reached adulthood, as do a very few individuals today, but all were social in their early lives, and the vast majority were social throughout their adult lives.

Survival

What does it mean to be "social"? On the simplest level it means that *humans need others for their very survival.* Infants need adults for their physical survival: for food, shelter, and protection. There is a great deal of evidence to suggest that infants also need adults for emotional support, affection, and love. Normal growth—even life itself—seems to depend on this support. Studies of infants brought up in nurseries with very little interaction with adults show us that these babies suffer physical, intellectual, and emotional harm and that this harm is lasting (Spitz, 1945). Of course, the horrible discovery in 1990 of infants brought up in Romanian government nurseries attests to the same problems: neglecting the basic emotional needs of children brings severe retardation of growth and often death.

Adults also need other people. We depend on others for our physical survival (to grow and transport our food, to provide shelter and clothing, to provide protection from enemies, and almost all the things we take for granted). As adults we also depend on others for love, support, meaning, and happiness. Human survival, therefore, is a social affair: almost all of our needs—physical and emotional— are met through interaction with others.

Learning How to Survive

To be social also means that much of what we become depends on socialization. *Socialization* is the process by which the various representatives of society—parents, teachers, political leaders, religious leaders, the news media—teach people the ways of society and, in so doing, form their basic qualities. Through socialization people learn the ways of society and internalize those ways—that is, make them their own.

Back to survival, for a moment: others are not only important for fulfilling our needs, but also for *teaching us how to survive.* We know how to do very little instinctively (suck, defecate, breathe, sweat, cry, see, hear, and other simple reflexes). But we are not born knowing how to deal with our environment. As newcomers we do not know how to get along in our world. We do not know how to deal with other people, weather, food sources, shelter, and so on. We do not know how to survive through instinct, necessitating our social nature. We do not have to learn that we need to eat; but we do learn how to get food (to grow it, hunt it, fish it, or buy it). In most societies (though not all) we must also learn how to build a shelter, use weapons, make clothing, and handle other people, to name only a few of the things that matter. In fact, we must learn thousands of things if we are to survive in the society we live in, from learning the ABCs to learning how to discourage others from robbing us to learning how best to dress and talk so we can be popular. In short, human beings live in a world where *socialization is necessary for survival.*

Individual Qualities

Besides showing us how to survive, *socialization is also necessary for creating our individual qualities.* Our talents, tastes, interests, values, personality traits, ideas, and morals are not qualities we have at birth but qualities we develop through socialization in the context of the family, the school, our peers, the community, and even the media. We become what we do because of a complex mixture of heredity and socialization. We may have certain biological predispositions, but how others act toward us, what they teach us, and the opportunities they provide for us are all important for what we become. As we interact with others, we choose the directions we will take in life: crime or legitimate business, school or on-the-job training, the single life or the married life, life on the farm or life in the city. Some of us may have all kinds of talent, but whether we direct it toward making money through selling illegal drugs or helping people solve their problems through psychoanalysis depends on our interactions and resulting socialization.

The treatment of women in our society highlights this point. In colonial days women were socialized to become the property of men. It was socialization not only by parents, neighbors, religion, and friends that accomplished this, but also limited opportunities for women in the larger society, and this prohibition told women *what* they must become to be useful in society. Eventually, women and men altered their relationship as women were increasingly socialized to take care of the household in return for male economic support. In the twentieth century, and especially after World War II, this relationship moved toward a more equal one. As economic opportunities opened up, women joined the paid labor force in great numbers. Their success in the political, educational, and economic worlds altered the expectations in society for women, and it increasingly altered the female role. After the war our view of the differences between women and men continued to blur. By the 1990s an acceptance had clearly evolved that women can do almost anything that has been traditionally reserved for men throughout our history. Such views influence the socialization of children; that socialization affects choices made in life. I never dreamed twenty years ago that women would ever compete in horse racing, bodybuilding, or fast-pitch softball. I never imagined that women would be successfully competing with men in the armed services, on police forces, and in business. My imagination was limited by my own socialization, which carefully distinguished what men could do from what women could do. Opportunity and socialization have influenced each other, and the result is a society less differentiated and stratified on the basis of gender. We are clearly living within a real-life experiment that offers clear support for the idea that socialization is very powerful for what people become!

It is important to see that socialization is very complex. It involves not only learning things but also modeling one's behavior on that of individuals whom one respects, being socialized by perceived opportunities "for people like us," and being influenced by one's successes and failures. When we see socialization this way, we can better understand the harmful effects of discrimination, segregation, and persecution. To be put down by others directly has an impact; to see others like oneself in a deprived existence has an effect on the value one places on oneself as well as the expectations that one develops for oneself. Of course, some individuals overcome such conditions, but these exceptions do not disprove the power of socialization. Indeed, they help clarify the importance of socialization as we try to identify the conditions that encourage individuals to be different from those around them. Socialization helps explain why poverty is so powerful a force on what children choose to do with their adult lives.

We can also turn this explanation around: the opportunities that wealthy and privileged people have in society socialize their children to seek directions closed to most other people in society: prestigious high schools and colleges, providing professional training that helps ensure high placement in society and a life of affluence. Robert Coles (1977) describes the final result of socialization in the wealthy class to be "entitlement": the children of the affluent learn that they are entitled to certain things in their lives that other children cannot take for granted and often do not even know exist. "The child has much, but wants and expects more, all assumed to be his or hers by right—at once a psychological and material inheritance that the world will provide" (p. 55). In what their parents give and teach, affluent children learn what they have a right to expect from life, what is their *due* because of who they are.

Socialization may not determine all that we are, but its influence cannot be easily denied. Much of what each of us has become can be traced to our interaction with others, and thus, our individual qualities are in this sense really *social ones.* The sociologist emphasizes how socialization influences our choices, abilities, interests, values, ideas, and perspective—in short, the directions we take in our lives. And, as we will see in later chapters, socialization is not something that happens to us in childhood alone; instead, it continues throughout our lives. At every stage we are being taught or shown by others how we should act, what we should think, and who we are. Early socialization may be the most important, but later socialization may reinforce these early directions or lead us in new ones. Socialization forms the individual actor and is the third way we are social beings: by our very nature.

Basic Human Qualities

We have looked at three ways in which we are social: our survival depends on others, we learn how to survive through what is taught to us by others, and we develop our individual qualities largely through socialization by others. A fourth quality of the human being attests to the importance of our social nature. Some religious and political leaders argue that human beings become human at the point of conception; others say that it happens at birth or after one year of survival. But in terms of certain *qualities that are central and characteristically human*—the use of symbols and the possession of self and mind—we are not human until we interact with others. These core qualities are socially created. We are, in a real sense, unfinished beings at birth. We have the potential to act as humans, but that potential is realized only through social interaction. Let us examine briefly each of these qualities.

The Use of Symbols   The more we understand about human beings, the more centrally important becomes their use of *symbols.* A symbol is something that stands for something else and that we use in place of that something else for purposes of communication. Although we communicate through the use of nonintentional body language, unconscious facial expressions, and so on, symbols have the additional quality of being understood by the user. Symbolic communication is meaningful: it represents something to the one who communicates as well as to the one receiving the communication.

Words are the very best example of symbols. They stand for whatever we decide they do. We use words intentionally to communicate something to others, and we use words to think with. Besides words, however, we also decide that certain acts are symbolic (shaking hands, kissing, raising a hand). And humans also designate certain objects to be symbolic: flags, rings, crosses, and hairstyles, for example. Such objects are not meaningful in themselves, but they are designated to be.

Where do such representations come from? It is true that many other animals communicate with one another: wagging tails, making gestures, giving off smells, and growling, for example. The vast majority of these behaviors, however, are instinctive. They are not learned, and they are universal to the species. They are performed by the organism automatically and usually do not appear to have any meaning to the user. (The bee, for example, will do its dance communicating to other bees where the nectar is located even when the hive has been emptied and there are no other bees around to see the dance.) The closer we get to the human being in the animal kingdom, however, the more the forms of communication take on a different quality: the acts represent something else only because it is agreed on in social interaction. In other words, *the tools of communication are socially based.* Because the meanings of symbols are socially based, what something represents is pointed out—intentionally taught—to the organism. Thus, the animal learns and *understands* that something stands for something else. When the act is performed, the animal does not simply give off communication but understands the meaning of it. It is clear that human beings depend on socially derived representations for almost everything they do and are; even if other organisms use symbols in this sense, the use is very limited.

This ability *to create and use symbols that are understood by the user is part of our social essence.* And this ability is so important to us that it probably qualifies as a central human quality alongside our social essence. Consider what we do with symbols: we use them *to communicate* ideas, feelings, intentions, identities; *to teach others* what we know; *to cooperate with others* in organization; and *to learn* cultural roles, ideas, values, rules, and morals. We hand down to future generations what we have learned, and they are able to build on what others have taught; symbols make *the accumulation of knowledge* possible. We use symbols *to think with*: to contemplate the future, apply the past, figure out solutions to problems, consider how our acts might be moral or immoral, generalize (about anything, such as all living things, all animals, or all human beings), and make subtle distinctions (between smart and not-so-smart candidates for office). Our whole lives are saturated with the use of symbols.

Selfhood   In a similar way, humans develop self-awareness only through interaction with others, and self-awareness, too, qualifies as a central human quality. Humans develop a realization that they exist as objects in the environment. "This is me." "I exist." "I live, and I will die." "I think, I act, and I am the object of other people's actions." This self-realization should not be taken for granted. It arises through the acts of others. We see ourselves through the eyes, words, and actions of others; it is clearly through socialization that we come to see ourselves as objects in the environment. Selfhood develops in stages, and each stage depends on a social context. Through interaction with significant others, we first come to be aware of the self, and we see it through the eyes of one other person at a time. (Children may see themselves through the eyes of their mother, then their father, then their nursery school teacher, then Mister Rogers—all in the same day.) Over time, our significant others merge into a whole, into "them," "society," "other people," or what George Herbert Mead calls a "generalized other," and we begin to use the generalized other to see and direct ourselves. We then see ourselves in relation to a group or society, in relation to many people simultaneously. We thus guide our own acts in line with an organized whole: our family, our elementary school, the United States, all people in our church, or all humanity. We see and understand a relationship between our acts and these other organized wholes.

Selfhood makes possible many human qualities, from the ability to assess our place in a situation to the ability to judge our own behavior to the ability to control our own behavior. We see and understand the effects of our own actions, and we see and understand the effects of the acts of others. We are thus able to plan strategy, alter our directions, and interpret situations as we act. For example, in choosing a major, students look at themselves: their abilities, interests, values, and achievements. They evaluate their experiences, future chances, and possible occupational opportunities. They will probably try to imagine what they would look like in a certain occupation and whether the work would be enjoyable. Selfhood also brings us the ability to judge ourselves: to like or dislike who we are or what we do, to feel proud or mortified. Related to this ability is what we normally call self-concept, identity, self-worth, and self-love. Selfhood also means *self-control,* our ability to direct our own actions. Self-control means that we can tell ourselves what to do. The more we investigate it, the more obvious it becomes that selfhood itself is a core human quality. And it is a *socially developed quality.* Without our dependence on interaction with others, selfhood would certainly not exist.

Mind   George Herbert Mead made sociologists aware that intimately related to selfhood and symbol use is the ability to think. Mead called this ability *mind.* Humans, like all other animals, are born with a brain, but the mind—the ability to think about our environment—is a socially created quality. Symbols are agreed-on representations that we use for communication. When we use them to communicate to our *self,* we call this thinking, and all this *communication* that we call thinking, Mead called mind. Humans do not simply respond to their environment; they point things out to themselves, manipulate the environment in their heads, imagine things that do not even exist in the physical world, consider options, rehearse their actions, and consider how others will act. (In other words, they figure out their world; they decide on how to act in situations; they do not simply respond to their environment.) This ability, so central to what humans are, is made possible through symbols and self, which (as we saw above) are possible only through social interaction.

To be social, therefore, means that humans need others to survive and need socialization to learn to survive. Socialization also creates our individual qualities. And social interaction is important for developing our essence: it creates our central qualities of symbol use, selfhood, and mind.

A Life of Interaction Within Society

Humans are social in a fifth sense, however. For whatever reason, *we live our entire lives interacting and embedded in society.* Observe our species: we are not simply around others all the time, we are doing things *with* others. Anyone watching human beings objectively should be amazed at how much their lives are affected by one another. We are constantly *social actors*: we impress others, communicate to others, escape others, con others, try to influence others, watch others entertain, display affection to others, play music or create art for others, and so on. Almost everything we do has an element of the social—it takes other people into account. As a result, we also end up *interacting* with others, and therefore, what we do affects what the others do. Action is built up back and forth as we do things together: cooperate, discuss, argue, teach, engage in conflict, play, make love, play tennis, or rear children. We are constantly involved in social action and social interaction, and this again is evidence of how important our social life is to what we are.

But we are also *embedded in social organization*. Our whole lives exist within groups, formal organizations, communities, and society. We live an organized existence, not an existence apart from others. Almost everyone spends his or her life in a world of *social rules* (morals, laws, customs) and *social patterns* (established systems of inequality, types of families, schools, and religious worship), a world that directs much of what he or she does. As we try to understand what human beings are objectively, we inevitably see animals who are born into a society they did not create, who are very likely to live their entire existence there, and who will find life filled with belonging to a host of groups, formal organizations, and one or a few communities. To observe humans in an environment that does not include a larger social organization is not to observe them as they actually live their lives. We are not solitary beings, but social ones; we exist within a social organization.

To emphasize the idea that human beings are social by their very nature is to see something very profound about what we are. Take away our social life and there is nothing left that we might call human. Our very survival depends on society; much of what we are both as individuals and as a species depends on socialization, and almost everything we do is based on and includes a strong element of social action, social interaction, and social organization.

Human Beings Are Cultural Beings

To say that human beings are cultural is to maintain that we are characterized by several other qualities not described above. Many animals are social, but what makes some animals cultural? The answer to this question entails determining what the foundation of a society is. Most social animals live together out of *instinct*. Nature commands that they cooperate, and it directs exactly how that cooperation should take place. Worker bees, queen bees, and other bees do not understand what they are doing, nor do they figure out how to play their various roles. Instead, they are born with instincts that control their behavior, making cooperation possible.

Some animals learn how to act in society, but much of that learning is *imitative*. They watch and do what others do. In this way they learn their place in the organization. In still other animal societies, adults actually teach the young what to do. This teaching is instinctive; that is, nature commands the organism how the young are to be trained. Now, it is difficult to determine how close to culture some animals come, but it is clear that human beings are cultural, and their social organization is founded on culture, not on instinct or on simple imitation.

As cultural beings, humans act in society as they do because they share a view of their environment. This shared view is sometimes called culture. *Culture* is a set of ideas, values, and norms (procedures, customs, laws, morals) that people use as a guide to understanding and self-control. It is how we are able to know how to act around one another in a cooperative manner. Humans discuss their world, learn about their world, and teach what they learn.

Knowledge is not lost with the individual organism but is passed down to others. There is a heritage that each individual within society learns and uses. People are not simply trained; with culture they are able to *understand* what they and others are doing and are supposed to do. Because of this cultural quality, societies differ considerably from one another. Each has a somewhat unique approach to living. Culture distinguishes organizations of people.

On the one hand, culture means that we see the world according to our social life; on the other hand, it means that we give meaning to our world. We do not merely respond to a world that acts as a stimulus on us. Instead, we understand it through the meanings that we learn in interaction. As our culture changes, so does our understanding of the world and our action in it.

Even our internal world is cultural, not simply physical. Our physical internal state may change as something happens to us (as someone points a gun at us or surprises us or tells us he or she loves us). But a change in our internal state does not automatically produce a response. Responses are defined, controlled, and directed by us, and they are guided by what our culture teaches. Between the internal physical response and what we do lies culture. Although many animals cry out toward their environment in what we might call "anger," human beings have the ability to understand that quality in themselves. They are taught by other people to distinguish anger from love, jealousy, pride, hatred, and fear. The culture that we learn tells us when it is appropriate to get angry and when it is appropriate to show it. We learn how to control anger, how to express anger, and how to feel sorry, guilty, or happy about our anger. It also teaches us many ideas about anger ("Anger is natural," "Anger is one important cause of prejudice," "The extent of anger is related to frustration"), and we apply these ideas to understanding our internal responses. Even the word *anger*—the label we give our internal state—is cultural. Experts are able to show us different types of anger and different levels of it. We can even learn when anger is "healthy" and "unhealthy," and we can learn how and when it can be "useful" or "harmful" to our goals.

We also label and act toward other people culturally, not "naturally." We see middle-class people and working-class people, conformists and nonconformists, nice people and nasty people. These labels are cultural. They help us divide up reality, and behavior that we perceive as deviant at one time or in one society may not be perceived that way in another (for example, polygamy, homosexuality, cocaine use, and divorce).

Max Weber through all of his work emphasizes the important point that we all live in a world of meaning. To understand human action, he argues, we must understand how people define their world, how they think about it. That thinking is anchored in a socially created culture. Weber focuses his attention on the influence of religious culture. He shows, for example, that in the seventeenth century, Protestantism was an important influence on the way people acted in the work world. In his view, Protestantism fostered a strong work ethic in society, encouraging individuals to strive for economic success. We are not isolated beings; through our social life we develop our thinking about the world, which, in turn, influences how we act in that world.

The Importance of It All

What difference does it really make that we are social and cultural beings? To be social and cultural means, first of all, that we are not set at birth but can become many different things and can go in many different directions. Because we are social and cultural, we are capable of becoming a saint or sinner, a warrior or business executive, a farmer or nurse. One can become only what one knows, and that depends on what one learns. Although biology may have something to do with differentiating us from one another, making it possible for some of us to excel in various spheres rather than others, our flexibility is still great, and thus society and socialization play an important role in what we become.

Societies vary greatly in what they emphasize and, thus, what they socialize their populations to become. We can become a peaceful people or a people who worship militarism. We can believe that the most important goal in life is to make money, or we can believe that the good life is one of unselfish giving. We can emphasize past, present, or future, people or things, competition or cooperation, this life or an afterlife, rock music or opera. Nature does not command what a society becomes; interaction and culture do, and thus we have evolved into many different societies. This means that as new circumstances and problems arise, people can reach new understandings and thus change their ways. It means that, in contrast to other primates, human beings are able to evaluate their ways and improve their cooperative endeavors.

To be social and cultural also means that to a great extent each of us is controlled by other people. The culture that we learn influences what we become and, in most cases, causes what we do. Most of who we are, what we think, and what we do can be traced to our social life. Unlike other animals, it is not nature that commands us. Nor, unlike what most of us think, is it free choice that characterizes many of our decisions.

To be social and cultural also alters our relationship with our environment. No longer do we simply respond to it in a fixed manner. Instead, we understand it through the eyes of a socially developed guide, and what we understand we use as a guide for self-control. There is stability in society over many years because culture is shared and embedded in each individual. There is change in society because as we interact we develop and teach new ideas, values, and norms. So, too, the individual: the individual remains the same or is changed depending on whether the culture changes.

Finally, to be social and cultural means that we gain certain qualities from interaction that make us *active* in relation to our environment, perhaps even free, to some extent. Our social life gives us the use of symbols, it develops selfhood in us, and it creates mind. We talk to ourselves about our environment. We figure it out. We solve problems. We make decisions and direct ourselves in that environment. In the end, we are able to control ourselves in ways that others do not plan for us and to develop ideas that are uniquely our own. In short, it is because we are both social and cultural that we are also able to think about the world and control what we do in it.

Summary and Conclusion

Look around you. Look in your classroom, on the campus mall, in your dorm, home, or apartment. Look at television, on the street, in the department store. Watch football games, symphony concerts, and serious drama. What is it that you see? What is the real essence of that being you see that we call human? The sociological answer is that you see

1. A being who is *social* in nature, who survives through a dependence on others, who learns how to survive from others, who develops both human qualities and individual qualities through socialization, and who lives life embedded in society.

2. A being who is *cultural* in nature, who interprets the world according to what he or she learns in society, and, therefore, a being whose nature is not fixed by biology but who is tremendously diverse.

It may eventually be found that alcoholism, homosexuality, intelligence, athletic skill, and so on have biological bases. It would be a mistake, however, to isolate and claim that it is only biology that matters. All of our qualities as individuals are encouraged or discouraged by society and its culture. Our rules, our ways of viewing others and ourselves, our rewards and punishments, and the expectations we have for ourselves and others are all social. In fact, it is critical to recognize that although biology may matter in explaining individual differences, it matters far less in explaining differences between groups of people. Groups of people differ primarily because of social and cultural differences.

REFERENCES

The following works attempt to explain the link between society and culture on the one hand, and the nature of the human being on the other.

Berger, Peter L., and Thomas Luckmann 1966 *The Social Construction of Reality*. New York: Doubleday.

Blumer, Herbert 1969 *Symbolic Interactionism: Perspective and Method*. Englewood Cliffs, NJ: Prentice-Hall.

Charon, Joel M. 1995 *Symbolic Interactionism: An Introduction, an Interpretation, an Integration*. 5th ed. Englewood Cliffs, NJ: Prentice-Hall.

Coles, Robert 1977 "Entitlement." *The Atlantic*, September.

Cooley, Charles Horton 1902 *Human Nature and the Social Order.* 1964 ed. New York: Schocken Books.

Cooley, Charles Horton 1909 *Social Organization.* 1962 ed. New York: Schocken Books.

Davis, Kingsley 1947 "Final Note on a Case of Extreme Isolation." *American Journal of Sociology,* 52:432–437.

Durkheim, Emile 1893 *The Division of Labor in Society.* 1964 ed. Trans. George Simpson. New York: Free Press.

Durkheim, Emile 1895 *The Rules of the Sociological Method.* 1964 ed. Trans. Sarah A. Solovay and John H. Mueller. New York: Free Press.

Durkheim, Emile 1915 *The Elementary Forms of Religious Life.* 1954 ed. Trans. Joseph Swain. New York: Free Press.

Elkin, Frederick, and Gerald Handel 1984 *The Child in Society: The Process of Socialization.* 4th ed. New York: Random House.

Erikson, Kai T. 1976 *Everything in Its Path.* New York: Simon and Schuster.

Freud, Sigmund 1930 *Civilization and Its Discontents.* 1953 ed. London: Hogarth Press.

Geertz, Clifford 1965 "The Impact of the Concept of Culture on the Concept of Man." In *New Views of the Nature of Man.* Ed. John R. Platt. Chicago: University of Chicago Press.

Gordon, Milton M. 1978 *Human Nature, Class, and Ethnicity.* New York: Oxford University Press.

Hertzler, Joyce O. 1965 *A Sociology of Language.* New York: Random House.

Lancaster, Jane Beckman 1975 *Primate Behavior and the Emergence of Human Culture.* New York: Holt, Rinehart and Winston.

Lane, Harlan 1976 *The Wild Boy of Aveyron.* Cambridge, MA: Harvard University Press.

Liebow, Elliot 1967 *Tally's Corner.* Boston: Little, Brown.

Lindesmith, Alfred R., Anselm L. Strauss, and Norman K. Denzin 1991 *Social Psychology.* 7th ed. Englewood Cliffs, NJ: Prentice-Hall.

MacIver, Robert M. 1931 *Society: Its Structure and Changes.* New York: Ray Long and Richard R. Smith.

McCall, George J., and J. L. Simmons 1978 *Identities and Interactions.* New York: Free Press.

Mead, George Herbert 1925 "The Genesis of the Self and Social Control." *International Journal of Ethics,* 35:251–277.

Mead, George Herbert 1934 *Mind, Self and Society.* Chicago: University of Chicago Press.

Nisbet, Robert 1953 *The Quest for Community.* New York: Oxford University Press.

Rosenberg, Morris 1979 *Conceiving the Self.* New York: Basic Books.

Shibutani, Tamotsu 1961 *Society and Personality: An Interactionist Approach to Social Psychology.* Englewood Cliffs, NJ: Prentice-Hall.

Shibutani, Tamotsu 1986 *Social Processes: An Introduction to Sociology.* Berkeley: University of California Press.

Spitz, R. A. 1945 "Hospitalism: An Inquiry into the Genesis of Psychiatric Conditions in Early Childhood." In *The Psychoanalytic Study of the Child.* Ed. Anna Freud et al. New York: International University Press.

Sumner, William Graham 1906 *Folkways.* 1940 ed. Boston: Ginn and Company.

Turnbull, Colin 1972 *The Mountain People.* New York: Simon and Schuster.

Warriner, Charles K. 1970 *The Emergence of Society.* Homewood, IL: Dorsey Press.

Weber, Max 1905 *The Protestant Ethic and the Spirit of Capitalism.* 1958 ed. Trans. and ed. Talcott Parsons. New York: Scribner's.

White, Leslie A. 1940 *The Science of Culture.* New York: Farrar, Straus and Giroux.

3

How Is Society Possible?

The Basis for Social Order

*I*n the fifteenth century Europeans went out into the world and conquered many peoples they found there. Over several hundred years (certainly through the 1950s) they destroyed whole societies and founded new ones. The Europeans also fought among themselves, drawing and redrawing boundaries between what they called nations and assuming that nations and societies were identical entities. In a sense they were playing with society, deciding that certain people belonged together in one society and that other people belonged in another society. Sometimes they drew lines, formed governments, made rules, and voilà—a new society was said to have been created. Civil wars were fought in many places (including the United States), and such wars altered what people thought constituted a society. Wars and peace conferences redrew lines. In 1871, for example, conquering Prussian armies created the German nation, a political entity, and it was declared to be one society. In 1918, after World War I, the victors redrew the lines of Germany, and the Austro-Hungarian Monarchy was broken into several nations, each assumed to be a distinct society: Austria, Czechoslovakia, Hungary, and so on. With the rise of Hitler, the German nation expanded to the borders preceding World War I, and the Nazis claimed that all Austrians and Germans living in what was then Czechoslovakia really belonged to German society. Others did not belong to German society; they would be exterminated, or they would serve the German conquerers.

The confusion over what constitutes a society and how societies and nations differ is real and important. Societies have existed from the beginning of human life. *A society is simply an organization of people who share a history, a culture, usually a language, and an identity.* ("We are a society!" "This is my society!" "These are my people!") In contrast to other organizations of people, a society is usually thought to be the largest organization that one identifies with. In the modern world (since 1500) nations or nation-states were formed, consisting of people governed by a common law. *A nation is a political organization of people, including government, law, and physical boundaries.* The boundaries of nations are often close to encompassing a society; sometimes a nation is formed that includes several societies or a part of one society.

Is the nation of Lebanon a society? For a long time after World War II, Lebanon was a society that people could point to with pride as one in which many different peoples lived side by side. Both Christian and Muslim peoples seemed to be loyal to the same society, and there seemed to be a mutual respect. Even with their differences (economic, religious, and historical) a society had been created that seemed to work, one that most inhabitants seemed to feel loyal to and with a government that people seemed to obey willingly. Beginning in the 1970s, that society collapsed through civil war as Muslims fought against Christians, Muslims fought against Muslims, and Christians fought against Christians. Lebanon may exist on a map, but something happened that made it impossible for it to continue as one society. Indeed, maybe it was really never one society after all, but a nation forged out of many societies.

One of the greatest forces in the modern world is nationalism, or loyalty to nation. *Nationalism* means that a people feels that it constitutes a society and that it has a right to rule itself in its own nation. If no nation exists, as in the case of the Palestinians, then nationalism is a claim that one *should* exist. *Nationalism is really best understood as a feeling of loyalty to a society that either has become a nation with its own boundaries and government or has not yet become a nation but in which people feel that nationhood is their right.* Nationalism is a powerful force going back at least to the seventeenth century. In England, France, Spain, Austria, Prussia, Italy, and Russia were subjects loyal to their governments partly because those governments were thought to rule societies of which the people felt they were a part. The American Revolution in 1776 and the French Revolution in 1789 were major turning points for nationalism. Rulers who were perceived to represent only themselves and not the society were thrown out, and political organizations and leaders were installed representing the society to which people felt loyalty. Indeed, as Napoleon's armies conquered European societies and spread the ideals of the French Revolution, they unwittingly spread nationalism, and it was this feeling awakened in the hearts of many Europeans that eventually led to the expulsion of the French invader. These people, too, thought that they were part of a society and that another government had no right to rule them. And as England, Germany, France, the United States, and other Western countries created empires in the nineteenth century, they contributed to the spread of nationalism. Conquered peoples felt loyalty to their own society, and they claimed the right to create a government and boundaries that would be consistent with that society.

Nationalism is clearly one of the most important forces in the world of the 1990s. The Soviet Union—a nation created out of a mixture of war, economic unity, history, and force—is no more. It has broken into many nations, and societies long ago engulfed by the Russian empire are emerging again to claim that they, too, should become nations. The Middle East is filled with nationalistic fervor, threatening established political states, sometimes leading to civil wars, sometimes to wars among nations. A root cause of this turmoil is the search for a way to create a political order—a nation—for the society of Palestinians. Ireland, Germany, Israel, Vietnam, India, Pakistan, and Algeria are but a few modern nations whose history can only be understood in the context of nationalism in the twentieth century. Today we are witnessing an upheaval resulting, in part, from the world creating highly artificial boundaries and declaring, "Here is a nation. You people will have to learn to live together!"

Nationalism is a force that threatens the unity of every nation ever created. There is no end to it: all groups that consider themselves separate societies demand satisfaction. If the Nation of Islam (the Black Muslims) in the United States becomes its own nation, how about the kingdom of the Ku Klux Klan or the separate Native American societies? How far must the leaders of the former Soviet Union go to satisfy all the nationalistic tendencies within their society? Is the United Kingdom a society, or is it really a kingdom of England, Scotland, Wales, and Northern Ireland? Time may build a united society, but someday nationalism may again emerge and threaten to create several societies out of just one. Abraham Lincoln squarely faced the problem of nationalism in the South: he fought a war and forced the South to remain part of the nation of the United States. Instead of each state being its own nation, the United States was to be one nation. As difficult as it was (and sometimes is), the United States had an advantage: it was mostly a land of immigrants who voluntarily left societies in Europe, who fought and defeated the Native Americans, and who together created a new society and nation. Imagine the difficulties we would have faced if more distinct societies had evolved in our history. Allegiance to one nation would have been far more difficult, and strong movements toward separate nations would have been common.

Societies do not just happen. They are built, and they are built slowly. Societies face difficult problems, and unless they are dealt with, societies can collapse, face revolution or civil war, become vulnerable to outside forces, or be transformed into something very different from what anyone really wants.

It seems that most of us take society for granted. We are born into it, generally accept it, and believe that it will always be there, slightly different perhaps, but there nevertheless. Unless we seriously think about society, most of us rarely ask what makes it possible and what, if anything, threatens its existence.

Sociologist Georg Simmel put the question simply: "How is society possible?" This question goes back at least as far as seventeenth-century philosopher Thomas Hobbes, who asked, "How is order possible?" What makes possible our willingness to give up personal desires for the good of the whole? What factors go into the creation and the perpetuation of society? As much as any other question, this is the one that has inspired sociologists from the beginning; and, in discussions among thoughtful sociologists, this is the one that comes up time and time again. In some deep sense, human society seems almost impossible. Where it successfully survives, it seems almost miraculous.

Society Is Possible Through Social Interaction

What makes human society possible? At the heart of all society is interaction, people doing things with one another in mind: cooperating, communicating, sharing, arguing, negotiating, compromising, influencing. People have to interact for society to begin and for it to continue.

Interaction is the building block of society. Consider for a moment what interaction means (and it is not an easy concept to grasp). Interaction means that actors take account of one another when they act. I act with you in mind; you act with me in mind; I act with you in mind again. What I do at any one point depends on what you do, and vice versa. This is easy to see when we consider two people: I say hello to you; on hearing me, you say hello back; when you say hello back, I inform you that I'm depressed; when you hear that I am depressed, you ask me what's wrong. Back and forth we talk. Each of us reacts to the acts of the other, who may have initially reacted to our own acts.

Interaction is also easy to see in a group; for example, consider a football team. If we concentrate only on the eleven players, we see the quarterback telling the others the play, we see players altering their acts as they see what other players on their own team are doing (a guard misses a block, so a back picks up the block), a receiver goes out for a pass, and in what he does the quarterback sees an opportunity and throws a pass to that receiver. In the huddle the receiver declares to the quarterback: "Good pass."

Of course, we can also see that there is ongoing interaction between the teams on the field, and from a distance we can observe interaction among a number of teams. For example, because most of the teams play most of the other teams, we can declare that all the teams in the league interact. We can see coaches among the teams meeting and drawing up rules, and referees holding meetings to help ensure that the league has consistency.

It is harder to observe interaction in a larger area, such as a neighborhood, but it is there. Sidewalks, stores, street corners, playgrounds, and hundreds of other places are occasions for people to interact with one another. Everyone does not interact with everyone else at the same time, but if we observe carefully, we see a pattern of crisscrossing interaction among people within the area, which is more intense and continuous than that between those people and people outside the area. That is one reason we declare, "That's a neighborhood." We can say the same about the larger community: there is crisscrossing interaction within the community that is far more intense and continuous than the interaction with those outside.

Society, too, is defined in part by this interaction. When people from several communities interact on a continuous basis and when that interaction is far more intense and continuous within than with outsiders, we see the beginning of society. Look at the opposite position: when there is no interaction, there can be no society; when interaction is segregated into two or more distinct entities, we must say that there is more than one society among those people. This was the point of the Kerner Commission report on riots in the United States in 1960: America had become two societies, segregated, each with different problems, each with different interests. Whether we are one or two or three or more societies is debatable, but here I am only trying to make the point that to be a society there must be ongoing interaction.

Why is interaction important to society? In a large part, this is because *human interaction is symbolic.* Symbolic interaction means that people's actions are usually *meant to communicate* something to others, and that the others who are objects of the communication constantly try to *understand* the meaning of those actions. Interaction is not simply physical responses to stimuli. Because we intentionally communicate, individuals can share with others their interests, concerns, values, demands, ideas, intentions, and feelings. Because we try to understand what others communicate, we have an opportunity to learn something from others, leading either to disagreement or, more usually, to sharing. Ongoing interaction that involves intentional communication and understanding facilitates cooperation and the negotiation of disagreement, both essential for the development and continuation of society. The significance of symbolic communication cannot be understated:

1. Communication brings a means of knowing one another, making possible consideration of the other's needs and helping to ensure that one's own needs are expressed. It brings a process known as "taking the role of the other," understanding the world from the perspective of others in the situation.

2. Over time, communication makes possible "shared understanding" among people. This shared understanding includes a way of handling disagreements and compromising between people's various interests.

3. Communication brings a basis for continuing cooperation, a way of handling problems together as they arise.

4. Communication brings a means by which people new to the interaction can be socialized so they know how to act in the interaction.

5. Finally, communication lets people know when their acts are unacceptable. It is a means of telling others that they are breaking the rules, that they are not going by the established group procedures, or that their acts are wrong.

In each instance, symbolic communication contributes to the functioning of society. To be outside the communication channels of society (that is, to be segregated from interaction with others) is to be outside of society itself. If large numbers are outside that interaction—if they interact among themselves and are isolated from everyone else—the maintenance of the larger society is made more difficult.

Again, the significance of symbolic interaction is easiest to see in a dyad or small group, but here we will jump right into society. The United States exists as a society in part because people continuously interact (through travel, mail, telephones, computers, television, radio, newspapers, and business deals, for example). Through symbolic interaction I begin to understand the problems of the individual in the inner city, the lives of wealthy corporate executives, and the ideas of my political leaders. And through symbolic interaction with others I let them know my ideas, my interests, my values. Although I rarely agree perfectly with any of these people, over time an underlying agreement usually arises among us: poverty is a tragedy in American life, capitalism is a healthy American institution even though there are serious problems, a college education is a necessity. Sometimes continuous interaction will bring about serious disagreements among us, but more usually it brings understanding and some agreement. When there is continuous interaction over time, we come to "think like Americans," to adopt certain values (such as individualism), to believe certain core ideas (such as that time is money), and to accept certain customs and morals (Sunday is a day away from work, drug abuse is harmful, incest is wrong). This is what is meant by a people's *culture*, one of the other reasons that human society is able to exist. *Culture arises in symbolic interaction; it is learned from others in symbolic interaction; it disappears without symbolic interaction.*

Knowledge of others, being understood by others, and sharing culture are basic to all cooperation in society. Consider any service in society—medical care, television, distributing and selling goods, education. These work *because* each actor understands his or her role in relation to relevant others in society: the store owner understands what to do in relation to customers, potential customers, advertising agencies, distributors, wholesalers, producers, federal, state, and local governments, and so on. As infants are born into society and are socialized through interaction, they come to learn what to do within the cooperative order and what not to do. And as they violate society's rules, they are told through interaction (with parents, teachers, police officers, members of the clergy, or other representatives of society) that their acts are unacceptable.

Society Depends on Social Patterns

Almost every sociologist believes that as people interact, a set of social patterns develops among them and becomes an important influence over their actions. Indeed, these patterns distinguish a "bunch of people" from some form of organization such as "society."

A social pattern means that social interaction is regulated, a stability is established whereby individual actors know what they are to do in relation to the others. We might call this a *routine*, a body of rules, views, and grooves that help to ensure organized action among actors. People will develop many such patterns over a long period, and all of us are born into such patterns. Sociologists normally identify three important patterns: culture, social structure, and social institutions. Although patterns are developing all the time as we interact, we inherit those that are anchored in the past. For most of us, these inherited patterns are generally accepted as part of our taken-for-granted world. For society to exist, people cannot simply act according to impulse or according to what they feel at the moment. Any cooperative order demands a certain degree of self-control in line with the social patterns that exist. Representatives of society teach the individual these patterns and encourage—through both rewards and punishments—at least minimal conformity.

Culture

Culture is one of the social patterns in society. It is really made up of a whole set of smaller patterns we might call rules, beliefs, and values.

For there to be cooperation, there must be *rules*, and individuals must be willing to guide their actions according to these rules. Societies are guided by customs: for example, when to have sex, with whom to have it, how one should feel about it, how it should be done. Societies are guided by laws: how old one's sex partner may be, what gender one's sex partner must be, what relatives must be excluded from marriage, under what circumstances one must refrain from sex. Societies are guided by taboos (prohibitions with severe punishment): what relatives must be excluded as sexual partners. Societies are guided by morals: how many sex partners are right, whether it is right to have sex outside of marriage, and whether the individual has a moral obligation to respect the wishes of his or her partner. Societies are guided by procedures: the role of foreplay, the best ways to have intercourse, what to do after the sexual act is over.

Societies also have informal expectations: who should be assertive in the relationship, who should take the responsibility for unwanted pregnancies, and who should remain a virgin. All of these rules— customs, laws, morals, procedures, and informal expectations—matter to the individual and to society. They tell individuals how to act; they tell them how others expect them to act; they tell them how to expect others to act. They also work to control the individual; they help to ensure cooperation in interaction. In short, they aid society's continuation through regulating individual action according to rules that most people understand.

Besides rules, culture also includes *values* (what people are committed to, what they consider to be important in their lives), and agreement over values allows for more cooperative interaction. For example, a society may value materialism, individualism, and family life. These values influence action: they encourage people to work hard in order to make money for themselves and their immediate family. They encourage people to go to school to get an education in order to make money for themselves and their future family. Because these are shared social values, many people will take this same direction in society, facilitating cooperation. Without some shared sense of what is important, organization would become more tentative and less united, with individuals going in whatever direction they decided to go, and cooperation would be made far more difficult. Shared values make it much easier for people to understand one another's actions, again facilitating cooperation, because people know what to expect from others, and others know what to expect from them. Values are the standards we apply to specific situations. They guide what we choose to do. "They are unquestioned, self-justifying premises that account for much of the consistency in responses to recurrent situations among those who share a culture" (Shibutani, 1986:68).

Culture is also made up of a shared set of *beliefs*. People may believe that hard work leads to material success or that a college education leads to a good job. A common belief in American society today is that marriage leads to a fulfilling life. "The free market system is the most effective economic system" seems to be an important belief in American culture. We have also come to believe that "a good government is one that stays out of the affairs of the individual" and that "people can become anything they want." Such shared beliefs influence people's actions, and order and cooperation are made easier.

Marx saw through these patterns of culture. He maintained that a people's rules, values, and beliefs are exaggerations of reality and that there are generally understandable reasons why particular exaggerations occur. Much of culture, he wrote, is *ideology*, or ideas that act to defend society as it exists, including its inequality of power and privilege. An ideology is not created by all people in interaction; it tends to be created and expounded by those who have power in society. To say that culture binds society together meant to Marx that certain ideas are created by and for the powerful and that these ideas are taught to most people. These ideas are given the name *culture*, but in fact they are ideology. They do work to keep order in society, but they work because they defend the inequality that exists. Most sociologists would agree with Marx to some extent: if we examine culture carefully, we can see that the rules, values, and norms tend to be exaggerations that operate to protect the powerful in society and, in that way, help to establish social order.

Culture, then, means that people in society agree on many important matters—rules, values, and beliefs—and this agreement fosters the continuation of that society. Perfect agreement is far from possible or even desirable, but general agreement is not only possible but also of central importance to society. Individuals will always disagree and will interact to form their own culture within society. When disagreement becomes widespread, a serious challenge to the underlying culture arises, undermining one of the important bonds of society.

Social Structure

Social structure is another important social pattern that makes society possible. As people interact over time, they establish *relationships*, they *position* and *rank* themselves in relation to one another, and they learn and play *roles* in the interaction. Structure refers to regularities in the interaction, people acting in relation to one another according to established routines. Structure organizes people's actions in society. As in culture, people understand what others expect them to do, and they understand what others are supposed to do.

A social structure is a set of statuses (or what we also call "positions," "social locations," or "status positions") that arise in interaction. People actually fill these positions in relation to one another. They are students (in relation to teachers), members of the middle class (in relation to the working and upper classes), men (in relation to women), quarterbacks (in relation to the rest of the team and the coach), and first-year employees (in relation to old-timers and employers). There are thousands of positions in society. People come to learn what is supposed to be done in each position they enter or may enter, and together the positions create order out of what would otherwise be chaos.

Each position has a role (a set of expectations that other people have for action in that position), and the individual who assumes the position learns to play that role. Each position also has a perspective attached, and a certain degree of power, privilege, and prestige. Quarterbacks simply are expected to see the game differently from defensive ends, first-year employees do not see the business the same way their employers do, and middle-class people learn to see the world differently from working-class people. For good or for bad, social structure ranks everyone, placing each of us in a position higher than some and lower than others. It has a simple relationship to maintaining society: people are sorted, they are distributed throughout society, they learn appropriate behaviors and ways of thinking, and they learn to fit their actions into the whole complex system. Labor is divided. We each contribute to society in a small way; together whatever is necessary in society is accomplished. Each of us learns but a small part of the whole; yet each of us is able to do what has to be done for society to continue over time.

There is no claim here that people take on the rank they "deserve" or "earn." Societies differ in the extent that this happens. In almost all societies most positions are inherited directly or indirectly from parents. This system wastes much talent and therefore works against solving society's problems, and it causes anger among those who feel that the distribution is unjust. The important point, however, is that social structure also contributes to social order.

Structure aids society in a second way. It builds an interdependence among the actors and through this interdependence creates a commitment to the whole. Durkheim best describes this process. As we each do what we are supposed to do in our various positions, others become dependent on us. As we deal with others in their positions, we become dependent on them. I am dependent on students, the president of my college, and the state legislature, for example. They also need me to teach sociology. Of course, I am also dependent on people in the Fargo-Moorhead Symphony Orchestra, the Minnesota Vikings, and National Public Radio for my entertainment; on those working at Hornbacher's Grocery and Walgreen's Drugstore to provide many of my simple daily needs; and on those in the police department and courts to protect my family and me. This exchange of services—this mutual dependence—ties us all together, and each individual becomes more and more conscious of his or her place in the whole society. Indeed, out of this interdependence grows a recognition of a higher social morality that must prevail if our mutual services are going to continue. Thus, a common morality results, a tie to a moral whole: society. Durkheim writes that the "division of labor"—what I am here calling social structure—balances individual self-interest with a higher system of rules:

> We may say that what is moral is...everything that
> forces man to take account of other people, to regulate
> his actions by something other than the promptings of
> his own egoism, and the more numerous and strong
> these ties [the interdependence of positions] are, the
> more solid is the morality. (1893:331)

Structure and culture together help to ensure control over the individual, cooperation and interdependence among individuals, and order in society. They are social patterns developed over a long time and necessary for ongoing interaction.

Social Institutions •

Imagine that every society survives only if it is able to solve certain problems. Society, for example, must have ways to produce and distribute goods, control disruptive behavior, socialize the young, regulate sex, defend itself, carry on business with other societies, encourage the performance of all the necessary roles, and develop adequate means of transportation and communication. It must minimally satisfy the needs of its individual members. Sociologists have long debated what list of minimal needs is best. The point, however, is that each society develops its own ways to solve these problems so that society can continue. These ways are sometimes called *social institutions*. Institutions are the third set of social patterns that make society possible.

Socializing Institutions   How do we create willing, hardworking individuals who accept society's ways? We simply do not do this "naturally." Nature sets us up: we are helpless, so we must learn how to survive, and because we are all born into society, we learn that survival depends on the acceptance of society's social patterns. The socialization of the young is one central problem of all societies (see Chapter 2). What is the solution? Institutions. Families socialize us, as do the media, churches, and schools. Political institutions socialize us, as do the law, literature, and music. If successful, socialization causes us to internalize society. We become society; its rules become our rules; its truths become our truths. Of course, we do not know any better. We are confronted with a social structure and culture that seem almost natural to us because so many representatives of society teach us these patterns.

George Herbert Mead describes this process through his concepts "taking the role of the other" and "generalized other." As humans interact, they learn to take the role, or perspective, of others around them. First, they take the perspective of "significant others," individuals who are important to them. Eventually, however, as they grow, this ability to take the role of the other matures to include a "generalized other," the whole group, the society. Many individuals in the various institutions are joined in the individual's mind to become this generalized other. This process is necessary for the continuation of cooperation: as we take on the generalized other, we learn society's rules and internalize them. Socialization ultimately means that we develop the ability to control ourselves according to the rules of the group, and through this control we are able to take part in cooperative actions.

Integrating Institutions   Another problem that societies must face is *integration*. What holds the individual members together as one whole people? What institutions develop to meet this need? The public school and the family contribute to integration to some extent, but so do the law, the courts, and the prison system. Such institutions encourage conformity and punish nonconformity. Political leaders help bring us together, as do mass transportation and modern communication. Most voluntary organizations, from the Elks to the Boy Scouts to the Democratic party, are important for integration, because they bring the individual *into* society, control the individual, and attach the individual to the whole. And, of course, we cannot forget religious institutions. Religion reinforces society's rules, making those rules sacred. It creates rituals that uphold the rules and bring people together in affirmation of those rules.

Other Institutions   Societies must solve many other problems besides socialization and integration. According to one sociologist, Talcott Parsons, a society must also develop institutions to adapt successfully to its physical and social environment, to develop and work cooperatively toward goals, and to keep its population relatively satisfied with their lives. Societies develop political, economic, religious, legal, military, familial, educational, health, and recreational institutions to help ensure that such problems are handled. Government has to work; that is, it must efficiently achieve societal goals. It must arbitrate disputes and enforce rules to ensure social control. It must develop relations with other societies. It must provide welfare for those who are unable to care for themselves. The society's economic institutions must effectively produce and distribute goods; legal institutions must regulate people's activities; religious, educational, and familial institutions must help maintain individual satisfaction. Society works in part because it has developed such institutions, and, overall, they work.

Of course, that is not to say that the institutions that prevail are the only way to do something. Private health care has "worked" in U.S. society for a long time. Something else might have worked better, and perhaps the older our population gets, the greater will be the pressure to change to a different system. The debate over health care institutions is on, and by the year 2000 we will undoubtedly see profound changes in what we once fondly called "private health care." Thus, the institutions that develop depend on the society: its history, the wishes of the powerful, its culture, and its structure. A society's particular institutions (a public school system, separation of church and state, capitalism, democratic decision making, monogamous marriage, and the state prison system) are its ways of solving problems. Although they may seem the best to people in that society, there are always alternatives to these ways, and other societies may have developed the alternatives. What works for one society does not necessarily work for others. Some societies rely heavily on force; others are more dependent on socialization and self-control. However, there must be institutional patterns in all societies, and these must work to effectively solve the ongoing problems that every society faces.

Remember: society is created and maintained through symbolic interaction. It is also created and maintained through what happens in that interaction: the development of culture, social structure, and social institutions. Through these patterns choices are made for the individual, order and coordination of action are established in interaction, and the problems confronting society are worked out.

Society Is Made Possible Through Loyalty

Almost every definition of society ends with a recognition of the importance of *feeling*. Society exists in part because people feel something toward it. They feel loyalty and commitment to the whole. Of course, no society exists in which everyone feels that commitment; but without some commitment by large numbers of people, leaders of society are required to rely on force for conformity, and instability results.

Ferdinand Toennies, an important European sociologist writing in the nineteenth century, describes two types of societies, each based on a different kind of loyalty. In more traditional societies, commitment is based on a "feeling of community," an emotional bond in which the individual feels that he or she is part of something larger. A sense of "we" prevails, and a belief that my efforts are important not for me, but for *us*. In German, this is called *Gemeinschaft*, and Toennies's description is almost identical to the feeling of "we" that Charles Cooley identifies when he describes the primary group (a small, relatively permanent, intimate, and unspecialized group). Many gang members have this strong sense of loyalty, as do small religious and political groups, and some societies such as contemporary Iran, Nazi Germany, England in the nineteenth century, and Japan prior to World War II. Strong nationalism comes close to this feeling: "My society is very important to me. My life is part of it. I will defend it against all enemies. I get my importance as a human being in part because I belong to it."

In every society there will be people who feel a real sense of community and see themselves as part of something very great. Many Americans do; on occasion, almost all Americans do. Along with this sense of community, commitment to modern society is often characterized by a conditional loyalty. More thought is involved in the relationship: "I am loyal to certain principles, I give my commitment so long as society meets my needs. My society is important, but so am I as an individual." Thus, what enters into our feeling of commitment is a belief that society does, in fact, work in our interests. Loyalty becomes more conditional. Instead of *Gemeinschaft* (a sense of "community") we have *Gesellschaft* (an "association" of people), in which contract and reason are more prevalent. Instead of feeling part of a primary group in which a sense of "we" prevails, people feel part of a secondary group, in which a sense of "I" prevails, with loyalty depending on whether the society meets my needs.

All societies are a mixture of both kinds of loyalty, some able to get more emotional commitment, others relying more on conditional loyalty. In modern society conditional loyalty seems more prevalent. But in Nazi Germany we saw a modern example of strong emotional ties to society. As individualism increases, commitment to the whole becomes more difficult. As Erich Fromm emphasizes throughout all of his work, however, as individualism becomes unbearable, people often seek greater commitment to the whole.

Modern industrialized societies, such as the United States, face a dilemma that is impossible to ever resolve, but which is an important key to many of the problems that face the individual. How much loyalty to the whole? How much individualism and freedom? Nothing has been more important to my own life than seeking freedom. Yet can freedom exist among many people in a society where people are unwilling to give loyalty to the community and refuse to follow the social patterns even minimally? Can freedom exist without commitment to the whole? And if we give commitment to the whole, can we pursue our own dreams, develop our own ideas and morals? Durkheim writes over and over: can freedom exist without a shared agreement as to what is right and wrong?

Institutions, as I pointed out earlier, are responsible for dealing with the various problems society faces. One important task they perform is to try to create and maintain loyalty. Public schools, families, religion, and political leaders try to socialize us so that we *feel* good about being part of our society. In times of war the political leaders and the military work to bring a people together in commitment to society, and national symbols of all kinds—the flag, the president, the national anthem, the death of an important leader, for example—accomplish this purpose. Defining some people as outsiders, as deviant, serves to bring loyalty to society, as does punishment of those defined as outsiders. Rituals of all kinds help bring people together into a society, integrating them and causing them to feel a sense of belonging. Ritual is an action whose purpose is not purely instrumental (goal-directed), but that communicates something among people that is symbolic of "the whole." Ritual is social action, and its purpose is to bind people together and to bind them with the past. All the various institutions include rituals, which reaffirm, dramatize, and encourage loyalty to society (Wuthnow, 1987:140).

For there to be loyalty, institutions must also deliver. Government must prioritize goals and convince us that these goals are being achieved. Schools must teach, economic institutions must produce prosperity, employment, and a bright future, and the courts must justly punish. Religion and family must bring some meaning and security to individuals. For loyalty to exist in society, people must perceive that their society works, that the institutions do an adequate job in dealing with problems. This is especially important in modern society.

Earlier in this discussion I pointed out that without this feeling of loyalty and togetherness, society would be based primarily on force. Max Weber, in his brilliant analysis of authority, shows how relying on force alone brings insecurity to society. Too much effort has to go into surveillance of the population; fear is costly; and constantly punishing the population is a waste of talent. Loyalty brings a willingness to cooperate, conform, and obey. Weber shows that voluntary obedience is the basis for stable systems of power in society. He calls such systems "authority" and defines authority as "legitimate power." In short, loyalty to society brings a willingness to obey legitimate representatives of that society, so long as they, too, conform to the rules. A system of authority helps guarantee stability in society. Without it there would be a continuous refusal by the population to follow rules, which would cause leaders to turn to force. It is impossible to determine how many people in a given society recognize the system of power as legitimate, and it is also impossible to determine how many people must do so for society to continue, but any society is clearly at a great disadvantage if it is without a power structure that is considered legitimate in the eyes of a large portion of the population.

Conflict and Change Help Preserve Society

Human beings are not born ready to participate in society. They have no instinct that prepares them for their positions, the culture, and the institutions they find there. The particular ways of society—its institutions—are not developed out of an instinct. We must recognize that human society, instead, is a result of trial and error, continuous interaction, and socialization. A people interact, form social patterns, and use those particular social patterns in working out what they must do in their lives. And as they do so, they develop loyalty to that entity called society.

However, loyalty and order can be exaggerated. Society also depends on social conflict and social change. As Simmel emphasizes in his analysis, social conflict actually contributes to social stability. The problem for any society is to recognize that open conflict is healthy because it encourages change and aids leaders in dealing with serious problems. To suppress conflict stirs anger and eventually brings destructive social conflict, which, far from promoting the continuation of society, may threaten its existence.

Society also persists because it is able to change; that is, interacting people develop new ideas, values, and rules; they form new institutions; and they create a new social structure. It is impossible to objectively claim that a particular change is good or bad, because such claims depend on what one means by good or bad. But it is possible to claim that some change is necessary for stability. New problems inevitably arise, and new ways must be developed to deal with them. People interact and make new claims (for better wages, better representation, better living conditions), and if these claims are ignored, violent conflict may arise, leading to the paralysis or disintegration of society. New dangers to society—abuse of illegal drugs, a massive national debt, a permanent underclass—may arise, and unless they are dealt with through creative solutions, what we all have come to know as American society may be threatened.

What matters is that both conflict and change occur in such a way that a society is able to evolve rather than dissolve and that creative problem solving becomes more important than destructive violence.

Summary and Conclusion

Societies exist because of social interaction. Without interaction there is no society; with segregated interaction there are several separate societies. Social interaction simply means that people act with one another in mind. Social interaction is symbolic. People communicate. They understand one another. They share various aspects of the world they live in.

Societies exist because of what people share in symbolic interaction. Over time, people create social patterns in that interaction: culture, social structure, and institutions. Culture binds people together because they come to agree on several important matters: beliefs, values, and rules. Social structure distributes people in society, locates them, teaches each how to act in relation to the others, develops interdependence, and facilitates cooperation.

Institutions develop to solve society's problems. People are socialized through institutions, society is integrated, people are rewarded and punished, goods are produced and distributed, and goals are developed and worked for.

Societies are also able to exist because people feel loyalty. They feel that they belong. They are willing to obey those in positions of power because they regard them as legitimate representatives.

And finally, societies exist because they are able to change and respond to conflict. Their members solve problems rather than ignoring them; they devise creative solutions rather than trying old solutions over and over again.

We live in a complex world. It is sometimes difficult to know what has gone wrong in our society or in others. Examine the world carefully: you will be able to identify symbolic interaction, culture, social structure, social institutions, and feelings of loyalty scattered here and there. You will see that societies evolve new patterns that work to solve new problems. That will be a good beginning for unraveling the mystery introduced in this chapter: How is society possible? How is it able to exist?

REFERENCES

The following works examine the meaning of society and the general problem of social order. Each either makes a general philosophical examination of society or looks at the more specific ways in which the continuation of society is made possible.

Aberle, D. F., A. K. Cohen, A. K. Davis, M. J. Levy, Jr., and F. X. Sutton 1950 "The Functional Prerequisites of a Society." *Ethics, 60*: 100–111.

Ballantine, Jeanne H. 1989 *The Sociology of Education*. 2nd ed. Englewood Cliffs, NJ: Prentice-Hall.

Bellah, Robert N., Richard Madsen, William M. Sullivan, Ann Swidler, and Steven M. Tipton 1985 *Habits of the Heart: Individualism and Commitment in American Life*. New York: Harper and Row.

Berger, Peter 1963 *Invitation to Sociology*. New York: Doubleday.

Berger, Peter L., and Thomas Luckmann 1966 *The Social Construction of Reality*. Garden City, NY: Doubleday.

Blumer, Herbert 1962 "Society as Symbolic Interaction." In *Human Behavior and Social Processes*. Ed. Arnold Rose. Boston: Houghton Mifflin.

Blumer, Herbert 1969 *Symbolic Interactionism: Perspective and Method*. Englewood Cliffs, NJ: Prentice-Hall.

Charon, Joel M. 1995 *Symbolic Interactionism: An Introduction, an Interpretation, an Integration*. 5th ed. Englewood Cliffs, NJ: Prentice-Hall.

Cooley, Charles Horton 1902 *Human Nature and the Social Order*. 1964 ed. New York: Schocken Books.

Cooley, Charles Horton 1909 *Social Organization*. 1962 ed. New York: Schocken Books.

Coser, Lewis 1956 *The Functions of Social Conflict*. New York: Free Press.

Dahrendorf, Ralf 1958 "Toward a Theory of Social Conflict." *Journal of Conflict Resolution, 2*:170–183.

Dahrendorf, Ralf 1959 *Class and Class Conflict in Industrial Society*. Stanford, CA: Stanford University Press.

Davis, Kingsley 1949 *Human Society*. New York: Macmillan.

Denzin, Norman K. 1984 "Toward a Phenomenology of Domestic Family Violence." *American Journal of Sociology, 90*:483–513.

Deutsch, Morton 1973 *The Resolution of Conflict*. New Haven, CT: Yale University Press.

Durkheim, Emile 1893 *The Division of Labor in Society*. 1964 ed. Trans. George Simpson. New York: Free Press.

Durkheim, Emile 1895 *The Rules of the Sociological Method*. 1964 ed. Trans. Sarah A. Solovay and John H. Mueller. New York: Free Press.

Durkheim, Emile 1915 *The Elementary Forms of Religious Life.* 1954 ed. Trans. Joseph Swain. New York: Free Press.

Elkin, Frederick, and Gerald Handel 1984 *The Child in Society: The Process of Socialization.* 4th ed. New York: Random House.

Erikson, Kai T. 1966 *Wayward Puritans: A Study in the Sociology of Deviance.* New York: John Wiley.

Erikson, Kai T. 1976 *Everything in Its Path.* New York: Simon and Schuster.

Ewen, Stuart 1976 *Captains of Consciousness.* New York: McGraw-Hill.

Festinger, Leon 1956 *When Prophecy Fails.* Minneapolis: University of Minnesota Press.

Gamson, William A. 1968 *Power and Discontent.* Homewood, IL: Dorsey Press.

Geertz, Clifford 1965 "The Impact of the Concept of Culture on the Concept of Man." In *New Views of the Nature of Man.* Ed. John R. Platt. Chicago: University of Chicago Press.

Hertzler, Joyce O. 1965 *A Sociology of Language.* New York: Random House.

Lancaster, Jane Beckman 1975 *Primate Behavior and the Emergence of Human Culture.* New York: Holt, Rinehart and Winston.

Lenski, Gerhard 1987 *Human Societies: An Introduction to Macrosociology.* 5th ed. New York: McGraw-Hill.

Liebow, Elliot 1967 *Tally's Corner.* Boston: Little, Brown.

McCall, George J., and J. L. Simmons 1978 *Identities and Interactions.* New York: Free Press.

Mead, George Herbert 1925 "The Genesis of the Self and Social Control." *International Journal of Ethics,* 35:251–277.

Mead, George Herbert 1934 *Mind, Self and Society.* Chicago: University of Chicago Press.

Olsen, Marvin E. 1978 *The Process of Social Organization.* 2nd ed. New York: Holt, Rinehart and Winston.

Reiss, Ira L., and Gary R. Lee 1988 *Family Systems in America.* 4th ed. New York: Holt, Rinehart and Winston.

Rose, Peter I. (ed.) 1979 *Socialization and the Life Cycle.* New York: St. Martin's Press.

Shibutani, Tamotsu 1955 "Reference Groups as Perspectives." *American Journal of Sociology,* 60:562–569.

Shibutani, Tamotsu 1961 *Society and Personality: An Interactionist Approach to Social Psychology.* Englewood Cliffs, NJ: Prentice-Hall.

Shibutani, Tamotsu 1986 *Social Processes: An Introduction to Sociology.* Berkeley: University of California Press.

Shils, Edward S., and Morris Janowitz 1948 "Cohesion and Disintegration in the *Wehrmacht* in World War II." *Public Opinion Quarterly,* 12:280–294.

Simmel, Georg 1950 *The Sociology of Georg Simmel.* Ed. Kurt W. Wolff. New York: Free Press.

Skolnick, Arlene, and Jerome Skolnick 1986 *Family in Transition.* 5th ed. Boston: Little, Brown.

Strauss, Anselm L. 1978 *Negotiations: Contexts, Processes and Social Order.* San Francisco: Jossey-Bass.

Sumner, William Graham 1906 *Folkways.* 1940 ed. Boston: Ginn and Company.

Sykes, Gresham M., and Sheldon L. Messinger 1960 "The Inmate Social System." *Theoretical Studies in Social Organization of the Prison.* Pamphlet 15. New York: Social Science Research Council.

Szasz, Thomas 1986 *The Myth of Mental Illness.* Rev. ed. New York: Harper and Row.

Toennies, Ferdinand 1887 *Community and Society.* 1957 ed. Trans. and ed. Charles A. Loomis. East Lansing: Michigan State University Press.

Warriner, Charles K. 1970 *The Emergence of Society.* Homewood, IL: Dorsey Press.

Weber, Max 1905 *The Protestant Ethic and the Spirit of Capitalism.* 1958 ed. Trans. and ed. Talcott Parsons. New York: Scribner's.

White, Leslie A. 1940 *The Science of Culture.* New York: Farrar, Straus and Giroux.

Whyte, William F. 1949 "The Social Structure of the Restaurant." *American Sociological Review,* 54:302–310.

Wuthnow, Robert 1987 *Meaning and Moral Order: Explorations in Cultural Analysis.* Berkeley: University of California Press.

4 Why Are People Unequal in Society?

The Origin and Perpetuation of Social Inequality

Introduction

In the view of Voltaire, the French eighteenth-century philosopher, "It is because the very nature of society creates inequality that the purpose of government must be to work for equality." Voltaire yearned for the equality of all humans, but he recognized that his goal was difficult precisely because we all live in societies, and societies necessarily create inequality. Sociologists are generally in agreement with Voltaire, and they are driven to understand more completely why he is right.

Sociologists have been interested in inequality since the very beginning of their discipline. It was the central theme of all Marx's work. Much of what Max Weber examined involved inequality. Indeed, the works of almost every great sociologist contain an approach to the subject.

Most people ask questions about inequality, and people who seek a just world almost always identify it as a source of injustice. In fact, this problem probably brings more thinking and caring people to study sociology than any other problem.

Every time we have anything to do with other people, inequality emerges in some form or another. Take individual qualities to begin with. We are not all equally handsome, intelligent, outgoing, or talented in athletics. Or when we compare ourselves on more social qualities, others seem richer or more successful.

The United States is a society dedicated to a democratic ideal that includes equal opportunity for all. Yet if we are honest, we find it difficult to ignore the great inequalities that persist and cause us to fall short of our ideal. In 1990, for example, the wealthiest one-fifth of the population in the United States received 47 percent of the income, whereas the poorest three-fifths received 30 percent of all income (U.S. Bureau of the Census, 1990). In 1986, the wealthiest 10 percent owned 73 percent of all the wealth in the United States; in fact, the richest 1 percent of the population owned 42 percent of the total wealth (U.S. Congress, Joint Economic Committee, 1986). According to one estimate based on 1990 Federal Reserve data, the net worth of the top 1 percent of the U.S. population was greater than the net worth of the bottom 90 percent (Kennickell and Woodbum, 1992). And it seems in recent decades class differences have increased considerably: for example, between 1963 and 1983, the average wealth of the wealthiest 10 percent rose 147 percent, compared with 45 percent for the rest of the population (U.S. Congress, Joint Economic Committee, 1986).

Much of what sociology has done is to understand and document the various types of inequality. It is not an easy task. Weber writes that we are unequal in three orders, or social arenas: the economic order, the social order, and the political order. We might translate this into class (economic order); race, occupation, education, gender, and ethnic group membership (social order); and political position (political order). Furthermore, within every organization from university to place of employment we also find positions that are unequal. And whenever we interact in small groups or even with one other person, we develop a system in which our informal positions are unequal. There are leaders, followers, and sometimes scapegoats. Similarly, we find differences between organizations of people: management in a corporation is more or less powerful than a labor union. Harvard and Stanford have more prestige and privilege—sometimes even power—than most other universities. The National Rifle Association has great power in certain matters, as do the American Medical Association and the National Association of Manufacturers. Political parties are rich or poor. Some churches survive on few funds, whereas other churches are wealthy.

Why does such inequality exist? Is inequality inevitable in society? To answer these questions clearly, we will divide our study into two parts. First, we will examine the reasons why inequality arises in the first place. Second, we will explore the way in which inequality is perpetuated.

Of course, the timeworn explanation for inequality is "human nature." Sociologists do not normally use such an explanation, because it tends to be too simple and too difficult to prove, and it ignores how inequality arises in social interaction. Furthermore, even if human nature causes people to try to be better than others (although this is debatable, too), something else must be found to explain how a stable system of inequality arises and is maintained over time.

Sociologists generally claim that inequality exists because of *the nature of social organization.* Many things happen when social interaction occurs, and one of them is the emergence of a relatively stable system of inequality.

Why Does Inequality Emerge?

What happens in interaction that results in inequality? And, once created, why does inequality continue?

Marx is a good place to begin. He argues that inequality occurs for three basic reasons: division of labor, social conflict, and the existence of private property. First, we examine division of labor.

The Division of Labor and the Rise of Unequal Power

As a division of labor develops in society—that is, as people increasingly do *different* things—some of these activities give people advantages over other people. If one person farms and others do not, the others become dependent on that one. And if one person farms successfully (through skill, luck, exploitation, or cheating), that farmer is able to accumulate economic resources—for example, land, laborers, and capital—that give him or her advantages over those with fewer of these resources. More important, if the farmer employs workers (a division of labor), a system of inequality begins in earnest. To Marx, the employer exploits workers, gains at their expense, and becomes increasingly wealthy and powerful in relation to them. The employer can then consolidate a favored position, and advantages grow as they are used. Once the process begins in society, it cannot be easily stopped.

If everyone essentially performs the same tasks—that is, shares in labor equally—significant social inequality will not arise. If no

one is the employer and no one the employee, equality characterizes the society. The division of labor is a beginning to understanding the riddle.

Marx focuses on economic division of labor in the way we produce and distribute goods and services in society. But there also
seems to be inequality in our families, our friendship groups, and
our churches—indeed, in almost every organization. What brings
inequality generally? The answer remains *the division of labor*—between those who lead and those who follow. As an organization becomes large or complex (differentiated in functions), someone must
make sure that events go smoothly, take care of day-to-day decision
making, and guarantee that the organization works to achieve its
goals. An organization needs coordination, and this normally means
having a leader. But does leadership impose inequality? Robert
Michels (1876–1936) adamantly answers yes. Once the leader is
chosen (and it does not seem to matter how), certain forces are set
in motion giving the leader advantages over everyone else. The position of leadership in an organization gives the individual who fills
it more information about the organization, the right to make decisions on a day-to-day basis, and control over what others know.
Michels's pessimism concerning the possibility of equality in organization has been called "the Iron Law of Oligarchy."

Certain developments in organizations seem to increase the
division of labor. As numbers rise, for example, informal interaction
usually gives way to attempts by leaders to formally coordinate activities. In addition, conflict with other organizations encourages
centralization of power and division of labor as ways of surviving
and maintaining internal order.

Division of labor also occurs because of organizational complexity. The more different activities an organization—including a
society—performs, the more division of labor results. Tasks are divided, and new ranks in the organization are added.

This process is complicated, however. Although increasing the
division of labor normally creates greater inequality, at a certain
point other factors enter in and work for greater equality. Gerhard
Lenski, for example, points out that as society moves from an agrar-

ian to an industrialized stage—and thus to a greater division of labor—there is actually greater *equality*. This is because industrialization demands a more knowledgeable population, a more technically skilled work force. The elite needs more help in leading society, and part of the cost is sharing wealth and power. Lenski never claims that inequality ends, only that in industrialization inequality is tempered by a recognition by those at the top that to protect their position and for society to progress economically, there must be some sharing of wealth and power.

Often inequality exists because *we purposely create* a division of labor in which positions are unequal from the very beginning. Those who create organizations realize that success depends on a division of labor and the system of inequality. Anyone who enters that organization is placed in a rank. A new organization will set up a board of directors with a president, a vice president, and so on. A new restaurant will open with a manager, an assistant manager, a head waiter, and a chief bottle washer. Weber refers to such a system of inequality as "authority," the ranking of positions in terms of power, each one having a certain amount of legitimate power in relation to the others. Almost every society has purposely created systems of authority, and in modern society almost every organization does so as well. Every time we enter an organization, we fill a position within an authority structure and have more or less power than others in that structure.

Bureaucracy is an important example of a system of inequality that is purposely created. Weber saw bureaucracy as a system of organization that would come to dominate the twentieth century, and if we look around, we can see how right he was. Almost every large organization we enter is another example of bureaucracy (from the university to the hospital to the corporation). Weber's brilliant analysis of bureaucracy reminds us of some central points: (1) Bureaucracy is a system of organization set up *on purpose* to get things done as efficiently as possible. (2) Bureaucracy is a system of inequality created so that responsibilities and lines of authority are clearly distributed. Obedience to authority is a central value. (3) Bureaucracy is a form of organization whose inequality is regarded by actors as necessary and legitimate. It is a system of control carefully created to ensure that the commands of the few are carried out by the many. (4) Once formed, bureaucracy is almost impossible to dismantle. Why? Because the division of labor becomes a necessary tool to achieve the goals of the organization, and because those at the top have the means to control other people in ways unimagined in other types of organization.

Almost every classical sociologist underlines the importance of the division of labor for the creation of inequality. An organization in which there are no leaders or a society in which everyone does the same tasks seems like an impossible dream (or nightmare). The trick is to create a division of labor that has equal positions, and this seems to be impossible.

Social Conflict and the Rise of Unequal Power

Thus far we have focused on the division of labor as an explanation for inequality. Marx introduces us to a second source: *social conflict*.

Conflict means the struggle by actors over something of value. Where there is struggle, some actors win and some lose, or, in most cases, some simply get more of what they want than others. Conflict occurs when there is *scarcity*: not everyone can obtain what he or she wants because there is simply not enough to go around. Conflict also occurs when some people monopolize what is valued in society, and, as a result, others are denied. In either case—scarcity or unequal distribution—as some increase their possessions, others cannot.

Who wins in conflict? Victory is simply explained by power: those who win have more power over their opposition. This might be personal power (based on intelligence, strength, attractiveness, guns, or wealth, for example). Often it is an organization, group, or society that enters into conflict and wins through superior power (better efficiency, greater loyalty among members, wealth, weapons, or leadership, for example).

When people win in conflict, they are normally in a better position to continue achieving their goals. They are also better able to increase their advantage over others, which in turn allows them to add to their advantage and eventually create a *system of inequality*. This helps to ensure that they (and their group and their descendants) will continue in this advantaged position. Victory becomes institutionalized; that is, it becomes established in the way society operates. Those who win create a system that helps to guarantee their continued success. Thus, the Europeans come to America, conquer the Native American population, and establish a treaty and reservation system that guarantees continued ownership of the land and subservience by the Native American people. Two businesses compete for a market. One eventually moves ahead in that competition, and over time the one that is ahead attempts to protect that favored position through instituting a distribution system, a pricing system, and an advertising program that will continue its domination of the market. Normally, the result is a fairly permanent system of inequality between those businesses.

Building a permanent system of inequality is advantageous to the dominant group in several ways. One: it protects the advantages the group already has. Thus, a system of law and government, if it is heavily influenced by the rich and powerful, will help protect their property and their privileges against others in society. Two: it places the dominant group in a favorable competitive position for jobs, education, and housing. Three: it allows members of the elite to use those in lower positions as laborers, renters, and consumers, thus making life easier and increasing their wealth.

Imagine the world as a place of continuous social conflict: all individuals, groups, organizations, and societies are struggling for whatever is valued. As some win, others lose, and over time a fairly permanent system of inequality emerges. Some will be rich, some poor; some will be powerful, some powerless. An upper class emerges, and a poor class develops. Men rather than women come to control the economic and political order, and all kinds of laws, ideas, customs, and institutions arise that continue that control. Whites dominate nonwhites, and Protestants dominate Catholics. In each case a stable system of inequality is created that favors the powerful. This happens in every organization and group. Individuals engage in conflict in newly formed groups, from juries to clubs to study groups to families. In each case, as people win, they are able to establish ways of protecting their interests. In short, the victors in social conflict are normally able to ensure their favored position.

The clearest examples are instances of extreme domination. Until recently, South Africa has been characterized by laws, customs, and ideas that openly declare the separation of the races and exclude blacks from equal participation in society and from equal protection by the law. Such was also the case with American slavery. To different degrees, Christian, Jewish, and Muslim societies have always had a strict code governing the actions of men and women, a code in which women have been systematically excluded from full participation in the political, educational, religious, and economic orders. When the Nazis came to power in Germany, they made it illegal for Jewish people to hold decent jobs, passed laws stripping Jews of citizenship, and took over Jewish property. They eventually established an organized system of resettling the German Jews in concentration camps, where the vast majority were systematically murdered.

Donald Noel (1968) highlights the role of conflict in his theory of the origin of ethnic stratification. Three conditions are necessary. First, groups that have separate cultures and identities come together. Second, there is competition for a scarce resource, or there is an opportunity for the exploitation of one group by the other (both are examples of conflict). Third, one group has more power than the other and is able to exert itself successfully in the conflict.

As parties win in conflict, a system evolves that essentially perpetuates the resulting inequality:

Social conflict $\longrightarrow$ Triumph of a group $\longrightarrow$ Creation of a system of inequality that perpetuates the group's favored position

Two explanations of social inequality have been presented here: the division of labor and social conflict. As an organization or society divides tasks among its people, some will become more powerful than others. As people engage in social conflict, some will win and some will lose, and over time a system of unequal power will be created. There is no suggestion that those who win are evil and selfish (sometimes they are and sometimes they are not), but most of us who succeed will be motivated to preserve the kind of world within which we were successful and will do things to protect that world.

So far we have linked social conflict to social power: social power allows some to win in social conflict; winning in social conflict, in turn, generally brings greater social power. We have also linked the division of labor to social power: once a division of labor is created, some positions emerge with more power than the others; positions with more power are able to ensure the continuation of a division of labor that favors them.

The easiest way to understand the meaning of *social power* is to recognize that it enables actors or organizations to achieve their goals in relation to others. To have a lot of power means that one is able to triumph over others. *Because of both social conflict and a division of labor, a system of power inequality arises in all groups, organizations, and societies.*

The Rise of Private Property and Unequal Privilege

The division of labor and social conflict not only create power inequality but also distribute *privileges* in an unequal manner. To understand this, let us consider social conflict once more.

Social conflict means that actors struggle for something: some win, some lose. There is struggle over power, but there is also struggle over other items that are valued by the actors: land, money, business, good housing, good education, safety. There is struggle because these items are *scarce*; the more some people have, the less others can have. If there were enough for everyone's needs, there would be no need for more, and there would be far less conflict; if such things were equally distributed somehow, there also would be far less conflict. However (as Marx reminds us), equal distribution is almost impossible when *private property* is established. Then people are able to own what they can get, and accumulation of what they own becomes difficult to limit. Jean-Jacques Rousseau describes this fact most clearly:

> The first man who, having enclosed a piece of ground,
> bethought himself of saying, "THIS IS MINE" and found
> people simple enough to believe him, was the real
> founder of civil society. From how many crimes, wars
> and murders, from how many horrors and misfortunes
> might not anyone have saved mankind, by pulling up
> the stakes, filling up the ditch, and crying to his fellows,
> "Beware that the fruits of the earth belong to us all, and
> the earth itself to nobody." (1755:207)

If everyone owned everything, if no one had more of a right to material things than anyone else, inequality would be insignificant. In
most societies, however, private property exists and is cherished.
Some people, in conflict over material things, win and accumulate
slaves or land or money. It becomes *theirs*. The more of such things
some people acquire relative to others, the greater is the inequality
between them. Weber calls such things privileges. *Privileges* include
all the benefits the actor receives because of his or her position in
society or a social organization—for example, income, housing, office space, health care, and opportunities for education. We become
unequal, therefore, in both power (our ability to achieve our will)
and privileges (the benefits we receive), and both inequalities arise
from social conflict.

An inequality of privilege results not only from conflict but
also from the division of labor. Employers make more money than
employees; doctors make more than nurses; rock stars make more
than teachers. Why should this be? Obviously, those who gain powerful positions in the division of labor are in the best position to increase their privilege. Owners of factories have more power than
others—and therefore more opportunity—to increase what they receive in society. Those who have less powerful positions can increase
their privileges only through organizing and taking a bigger share
from those above them.

Another reason the division of labor leads to the unequal distribution of privilege has something to do with market conditions:
positions are given different amounts of privilege in an organization
because of a combination of (1) importance to that organization (the

most important positions tend to get the most privileges), (2) the amount of training and sacrifice one must go through to prepare for those positions (the more the training and sacrifice, the greater the privileges), and (3) the scarcity of people for those positions (the fewer the people seeking the position, the greater the privileges). Conversely, any position that is not very important, takes little training, and has many people competing for it will be paid much less. After all, positions that are essential must be filled; to fill them, people must be attracted; to attract them, privileges must be used. Market conditions and social power combine to help ensure that the division of labor creates a system of unequal privilege.

However, our discussion becomes even more complicated. Not only does social power bring more privilege, but also the reverse is true. Social privileges (for example, money, land, factories) also bring greater social power. Power depends on many things, but wealth is certainly among the most important. Those in well-paid occupations have more power than those in low-paid occupations. Those with wealth influence what happens in the economy, in government, and in the media. Therefore, it is important to see the relationship between power and privilege: power begets privilege, privilege begets power. Both are the result of social conflict and division of labor. It may be helpful to illustrate these complex links:

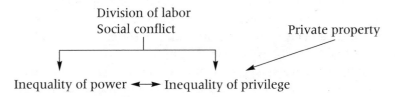

The division of labor differentiates people and creates a system of inequality in both power and privilege. Social conflict creates winners and losers; the winners are then in a position to increase both their power and privilege. Social conflict and the right to private property create unequal privileges. Unequal privileges, in turn, create unequal power, and unequal power allows for increasing privileges. Employers tend to be high on both power and privilege; employees tend to be low on both. Leaders tend to be high on both;

followers low. The rich tend to be high on both; the poor tend to be low. Whites, corporate executives, and army generals tend to be high on both; nonwhites, janitors, and privates tend to be low.

Prestige and Inequality

Another type of inequality emerges in this complex process, and this inequality influences both power and privilege. This is social *prestige.*

Marx did not emphasize prestige in his analysis, but other sociologists, such as Weber, did. Prestige has to do with how other people evaluate us. All of us wish to be respected by others. Some of us work our whole lives to gain and keep this respect. Respect is personal. It centers on the individual qualities that a person possesses: integrity, intelligence, talent (musical, athletic, artistic). "Man, do I respect that guy!" Prestige is like respect in that it has to do with how others view us; however, prestige is *social.* It is honor that others accord an individual because of the *social position* that he or she has in society.

As positions become differentiated in society or in any social organization, people come to attribute different degrees of honor to those positions. Individuals are judged on the basis of *where they are* in the social structure: are they men or women, white or nonwhite, professional or blue-collar worker, secretary to the president or secretary to the vice president. People accord high prestige to the top executive and dishonor to the prostitute in society. Officers in the army have more prestige than the enlisted personnel, professors have more prestige than instructors, and the rich more prestige than the poor. It is not always clear why some positions gain more prestige than others, but in large part it is because of the *power* and *privilege* that go with those positions. Powerful and well-paid positions tend to be the most honored (the upper class, the chairman of the board, the movie producer, members of Congress). Those without power or privilege are normally low in prestige (the poor, the unemployed, the unskilled, minorities).

Not only do power and privilege influence prestige, but also prestige influences power and privilege. Prestige is one basis for power: one can use prestige to achieve one's will. Someone, for example, who has high prestige in any authority structure or who is rich, white, male, and so on has an important advantage both in interaction with others and in the general society in achieving his or her own goals. Prestige can also influence privilege: those honored in organizations are normally given (or simply demand) special treatment. All three qualities are strongly linked: prestige begets power and privilege; power begets prestige and privilege; privilege begets prestige and power. Together these benefits help create a ranking system in all organized life.

In a sense our picture is complete, and we can begin to see why inequality arises in the first place: *A system of inequality arises in society because of a combination of division of labor, social conflict, and private property. The resulting positions develop different levels of power, privilege, and prestige. Power brings privilege and prestige; privilege brings power and prestige; prestige brings power and privilege.*

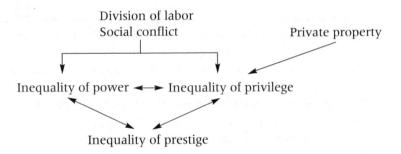

Thus, all of us are born into a society with an established system of inequality that has developed from a combination of the division of labor, social conflict, and private property. We enter organizations—schools, businesses, professional associations—each of which has a system of inequality (usually intentionally created) with positions in place. Although each actor has some leeway in these positions, each is also faced with the fact that positions have over time developed more or less power, privilege, and prestige than other positions. And if we interact and form our own organization, social conflict, division of labor, and the right to private property will ensure that a system of inequality will arise (even if we do not intentionally create it).

Why Does Inequality Continue?

Once a system of inequality has been established, it is difficult to alter. Of course, it changes slightly over time, but it tends to perpetuate itself. It seems that five mechanisms work to cause this stability:

1. Efforts of the powerful

2. Social institutions

3. Culture

4. Socialization

5. Instruments of force

Let us examine each in turn.

Efforts of the Powerful

Marx and Michels (the "iron law of oligarchy") explain how the powerful protect the system of inequality. When one gains a high position, one has the resources to protect oneself. Those who are not favored within the system have fewer resources to protect themselves and have little ability to change a system that keeps them low. In short, *inequality is perpetuated through the power of those who benefit from it.*

Marx makes a simple and convincing argument. When some people own the means of production, he argues, they have great power, and they use that power to protect their position and to increase their wealth. Thus, there is a strong tendency for the rich to get richer and to effectively protect their positions in society.

Those who own the means of production are called appropriately by Marx the "ruling class." Control over large businesses gives them control over people's jobs, the communities people live in, the products that are made, the economic decisions that affect the society, even the world. Control means that any decisions made will probably help the rich and powerful.

Marx goes much further, however. Control over the means of production—economic power—is translated into other types of power. Economic power influences government: the rules government goes by, the people who fill its positions, and the laws it makes. The ruling class influences media, the schools, the courts, and almost every other sector of society.

Why does this happen? Simply put, it is in everyone's interests to influence successfully the direction of society. I want to—so do you. But I will try to influence it differently from you. For example, you will try to lower tuition, but I will try to raise faculty salaries. Feminists, African Americans, lawyers, unions, ministers—to name but a few—all have their own agendas, and all would like to see their needs met. The rich and powerful have interests too, but the difference between them and everyone else is that they have greater resources to use in ensuring that their needs are, in fact, met. Thus, although the society never works completely in their interests, the tendency is for it to work in ways that are consistent with what the powerful want.

Remember: The question we are considering is how inequality is *perpetuated* over time. We now have one answer: Inequality is perpetuated because those who are wealthy and powerful are in the best position to ensure that their interests are met throughout society.

Michels adds to our understanding of this process. He emphasizes the political side of the coin. Michels is interested in what happens to a group or society that chooses leaders; he is less interested in economic power. Michels simply makes the point that leaders in any organization will eventually develop different interests from everyone else in the organization. Once selected, democratically or otherwise, leaders come to regard their positions as "theirs," and they tend to institute policies and pursue goals that are consistent with that belief. Leaders eventually unite and form a self-supporting group that distinguishes them from everyone else. One of the leaders' primary goals becomes perpetuation of their favored position.

To some extent Marx and Michels agree: those who have power (economic or political) develop interests different from everyone else's (to maintain the inequality that exists), and they are in the very best position to influence society to work in these interests. We must therefore begin to understand the perpetuation of inequality by recognizing that it is in the interests of the most powerful to do what they can to maintain the system that favors them, and their power gives them the ability to do so.

Prevailing Social Institutions

The result of this control by the ruling class is the creation of *institutions*, the ongoing and legitimate ways of doing things in society. Institutions, as we saw in Chapter 3, are the established procedures that help to ensure the continuation of society, and to some extent they benefit most people in society. However, they normally benefit the wealthy and powerful the most, for they are created and supported by them, and they generally work to protect the system of inequality that exists.

Over time, for example, political institutions are created in society to pass laws and carry out those laws. The United States has separate legislative, executive, and judicial branches of government to do this. It is characterized by a two-party system, an electoral college, federalism, separation of powers, and civilian control over the military. These are our political ways. We also have economic institutions such as multinational corporations, a Federal Reserve system, a stock market, private property, and private enterprise. We also have educational, religious, health care, military, kinship, and entertainment institutions. The United States as a society works—or does not work—because of its institutions. One of the primary reasons that the Soviet Union was forced to turn from a communist system was simply because its institutions could not solve the problems it faced in the 1990s.

It is important to realize that institutions generally work for the society as it is. If they seem to work, they continue; if they seem to work in the interests of the powerful, they are especially encouraged. *Once a society develops a system of inequality, the prevailing institutions tend to work in such a way that the inequality is maintained or even increased.* It is easier to see this process in other societies than in our own. Saudi Arabia is a society where almost everything that exists works to maintain the wealth and power of a few families and the dominance of men over women. That is the way the government, the economy, the religion, the military, and the family work in that society. Apartheid in South Africa, the forced separation of the races for purposes of domination by the whites, has been maintained through a complex set of institutions. In China, government, military, education, and media combine to help ensure the continued dictatorship of a small party.

Unless equality is a value that a society truly pursues, the institutions will normally protect and expand inequality. Poverty continues in the United States because institutions are not set up truly to deal with this problem. Our tax system does not substantially redistribute wealth, it does little to effectively limit the wealth that one can achieve, and in the past decade it has actually contributed to greater inequality. Our schools, government, welfare system, and economic system may be wonderful in some ways, but they tend to protect the system of inequality that prevails and to maintain people in the class positions of their birth. Institutions maintain our segregated society, they generally support the inequality between men and women, and they protect the power and privileges of a political and economic elite. This is why sociologists tend to see the perpetuation of inequality built into society itself. It is rare to see a society whose real purpose is to maintain a system of equality.

Remember Voltaire's warning described at the beginning of this chapter. Unless a society really makes efforts to create and maintain equality, the tendency will be toward a state of inequality. It is easy to see why this is true when we recognize the tendency for social institutions to protect and expand social inequality.

Thus, the second reason why inequality becomes perpetuated in society is that institutions generally work in that direction, partly because the powerful have the greatest impact on the nature of those institutions.

Culture: The Acceptance of Inequality

Over time, most people come to accept the inequality that exists. Obviously, this acceptance aids its perpetuation.

It is important to remember that people are socialized into the society within which they are born. As we saw in Chapter 2, socialization means that we take on many of the qualities that people who represent society (parents, political leaders, religious leaders, media leaders, and teachers) teach us. To a great extent, we take on *their* language, *their* rules, *their* values, *their* expectations. This is culture.

Culture almost always includes justifications for inequality. For example, Americans generally believe that we live within an economic order in which people will be justly rewarded for hard work: "If you work hard, you can rise to the top." In a sense, we believe that the system of inequality is somehow just and democratic, rewarding those who ought to be rewarded. This is the type of cultural idea an individual comes to believe through socialization. Peter Berger argues that most societies actually develop two ideologies that serve to protect inequality. One ideology legitimates the position of the upper classes, usually arguing that these people are somehow superior or more deserving. (For example, they are more talented, more hardworking, or superior in culture.) The other ideology argues that poverty is a consequence of sin or laziness and that good behavior by the poor will eventually be rewarded in the afterlife or in the next life.

In European societies, inequality was justified for centuries through arguing that it was God's plan. God chose the rulers and also favored an upper class whose purpose was to lead the masses. Although revolution eventually destroyed this idea, it held on for a long time. Indeed, in much of the world today inequality is still seen to be God's will, and in most cases people are taught to spend their time and energy doing other things besides trying to change society to make it more equal.

For a long time, people in the United States denied the existence of either a rich upper class or a class of poor people: "We are the land of equal opportunity for all." To believe this is to deny the effects of inequality, and such an idea works well to protect the inequality that actually exists. It is difficult to identify exactly what American culture consists of, but try to list the very basic ideas, values, and morals that we believe in, and you will see that they work to uphold the system of inequality that prevails: "The poor do not really want to work [and thus deserve their fate]." "Capitalism with little government regulation and taxes is the most just and efficient economic system." "People have a right to make what they can, keep all that they make, and pass down to their children all that they keep." "We may be a class society, but everyone has the chance to strike it rich." "People are naturally competitive, selfish, and lazy.

We all work to beat the other guy, we all take whatever we can get, and we have to be paid well if we're going to work."

Or consider the ideas that accompany other kinds of inequality: "Women are not naturally capable of competing in the political and economic world." "God did not mean for women to work outside the home." "Women have a moral obligation to obey their husbands." "Blacks are naturally inferior to whites." "God meant for blacks to obey whites." All of these ideas—and many more—have existed in our own society and have worked to retain systems of inequality. Similar ideas exist in other societies. In India people are born into castes that do not change. How is this system justified? One's position in the next life depends on how well one accepts his or her position in this lifetime. Caste is a test; acceptance of position is a moral virtue. And when we become too sophisticated to believe old ideas, we develop new ones to justify inequality: "Women are naturally different from men and will do better than men in some things, such as rearing children, and not in others, such as mathematics." "I believe that blacks should be equal, but the fact that they are not is their own fault." "It is morally right for us to dominate them because of their cultural inferiority."

In *Oppression*, Turner, Singleton, and Musick carefully show how the dominant beliefs in the United States concerning African Americans have changed as the relationship between blacks and whites has changed (1984:170–176). We always work out a new ideology that justifies inequality. Before 1820, whites described blacks as uncivilized heathens, the curse of God, and ill-suited for freedom. After 1820 and before the Civil War, slavery was justified as good for both blacks and whites, an institution that civilized and protected African Americans. After the Civil War and before World War I, while all areas of life became segregated, blacks were described as inherently inferior, and thus segregation became necessary for the protection of whites. From World War I to 1941, as African Americans moved north, black inferiority became a "scientific fact," and segregation was described as natural, distinctive, and desired by both races. After World War II, as discrimination and segregation were increasingly attacked, ideas favoring inequality were fewer. After 1968, however, inequality again became justified: there

is black inferiority, and it is due to the African Americans themselves, especially their "lack of motivation." Many Americans have
recently come to believe that we have moved too fast toward equal
rights and that society has done enough.

Culture is also important in what it does not teach. The 1990s
have seen the collapse of the Soviet domination of Eastern Europe
and the rapid decline in communist institutions within the Soviet
Union. The U.S. media have responded by paying tribute to certain
American institutions and values, especially private property, a free
market, and freedom to own one's own business. Somewhere along
the line these values have come to dominate our culture more than
equal opportunity, social justice, respect for the individual, and cultural pluralism. It is not that we no longer believe in the latter; it is
simply that the former values have come to dominate our thinking.
By ignoring equality and social justice, culture tends to support inequality and lack of social justice. It does not encourage people to
question the extent to which they exist in society.

Even in organizations we see ideas that serve to justify inequality there: "The leader knows what is going on; the rest of us
just don't have enough information to make intelligent decisions."
"Democracy, where everyone votes, is inefficient." "The real world
isn't equal. Why should this organization practice equality? We have
to run this place like a business, don't we?"

The key point should not be missed: society—all social organization—develops a system of belief ("culture"), and that system
of belief (among other things) functions to protect the inequality
that prevails.

But where does the culture come from? As we saw in Chapter
3, a culture is a set of ideas, values, and rules that people adopt over
time as they deal with their environment. These ideas, values, and
rules are partly true, but more important, they are *useful*. Our culture works for us; other people's culture works for them. Culture
explains and guides our actions. But who is *us*? Whom does culture
really benefit? In a way all people within society benefit from their
culture. It provides them with guides to living. However, culture
most certainly benefits those people who gain the most from society: those at the very top of society.

It should come as no surprise that those at the top try to ensure that their ideas, values, and rules prevail in society. Think of
the society giving rise to many ideas. Which ones are believed, and
which ones are rejected? This question is not answered simply, but
be aware that ideas, values, and rules must have sponsors, groups
that push for their acceptance. Those groups with the most power
will have the best chance for their ideas to be accepted. To some extent Marx is correct: the rich not only produce the goods of society,
but also to a great extent they produce (and spread) the society's
culture. And, of course, when ideas, values, and rules are put forth
that are not consistent with those of the elite, they will be opposed
and have less chance for acceptance.

Inequality prevails in society, then, because it is supported by
culture; and culture, in turn, most reflects the ideas, values, and
norms of the most powerful in society.

Socialization: The Acceptance of Place

Besides the fact that people are socialized to accept the system of inequality itself, they are also socialized to accept their own position.
This is a complex process. We learn who we are early in life. Our
neighbors, parents, and teachers tell us in overt and covert ways our
ranks in society and what we have a right to expect from life: "People like us don't do those things." "Marry your own kind." "Go to
Harvard." "Be satisfied with the college in your own town." We are
taught what to expect from life if we work hard (generally slightly
above our present rank, whatever that may be), but rarely do we
expect great things without a realistic model to go by. Wealthy business executives socialize their children to expect wealth, fame, and
power. Lawyers socialize their children to expect a professional position. Of course, we are not simply the result of what our parents
expect, but they, together with teachers (who teach in class-based
schools) and friends (who tend to come from our class-based neighborhoods), show us where we are in society and teach us to expect
approximately that level.

In *Schooling in Capitalist America*, Bowles and Gintis highlight
how legitimating inequality and teaching people their positions is

an integral part of our educational system. Schools sort students into academic tracks, which distribute students into the occupational system and ultimately into the economic system. Schools teach discipline, hierarchy, and obedience, and they teach students to expect little control over their work. Working-class students learn obedience; upper-middle-class students learn leadership and innovation.

Or witness, too, how society socializes women to accept subordinate positions and how it tries to teach minorities to accept their positions. Society is not always successful, right? Some people try to make great leaps to overcome their position, and some succeed. Most do not succeed, not because of lack of effort or intelligence alone, but because real opportunity is denied by factors related to class and minority positions. And what happens if one tries and does not succeed? Well, this is one important way in which we are socialized: failure to improve our position causes us to change our sights.

Legitimate authority is one of the concepts that Weber introduces to sociology. Why is it that we willingly obey others over us? he asked. His answer is that most of us *believe* that they have a *right to command us*. Socialization induces people to feel a part of a community and to *feel an obligation* to obey the people who represent that community. We come to believe that we must accept the prevailing system of inequality as right if we are to exist as a community. A stable system of inequality is built into the community's tradition or law that loyal people feel an obligation to follow. Our system of inequality appears to most citizens to be *legitimate*, and most feel a moral obligation to obey those above them.

Thus, the system prevails through the collusion of the individuals who are socialized into society. Socialization brings the acceptance of a culture that justifies inequality, and it normally brings an acceptance of one's relative position in that system of inequality.

Society's Instruments of Force

Of course, some individuals refuse to accept their place in society and try to improve their position in any way they can, including going outside the law. Often, they realize that the system works against them and that to make it they cannot try the normal channels. Through their acts they threaten the legitimacy of the prevailing order. Procedures are instituted to discover, control, and punish such individuals. Police, courts, and prisons work to protect more than people; they also protect the system of inequality.

Some groups refuse to accept the system itself and organize to overthrow it. In our society such groups have some leeway: they usually have the right to say what they want and to write what they want. When they act outside the law to alter the system, we call them revolutionaries, and we use force to stop them. All societies draw lines, and all try to control groups with force if they threaten the established order, including the prevailing system of inequality.

Although crime and revolution upset order and create hardship for all people in society, those at the top of the system of inequality have the most to lose. It is their favored status that is most at risk. They have the most property to lose. The advantages they enjoy are threatened. It is vital to them that what they worked for or inherited be protected. They therefore take an active interest in politics, law, and law enforcement. Marx goes so far as to say that the state's purpose is to protect the ruling class. At the very least, we can see that legitimate instruments of force are important ways in which those at the top are able to protect themselves and maintain the system of inequality.

Every organization establishes instruments of social control that protect the structure. Colleges use grades, threats of suspension, and, sometimes, refusals to cooperate with threatening students. A business can fire, demote, or refuse to promote. A family can send the offender to his or her room or "ground" the individual for two weeks. An informal group can simply let the member know that he or she is not liked and may not invite the individual to future activities. Of course, there is some individuality and some freedom; but threats to the prevailing system of inequality are always reacted to.

Summary and Conclusion

Inequality arises from social conflict, the division of labor, and the existence of private property. It also arises from the mutual influence of power, privilege, and prestige.

Inequality continues over time. The efforts of those at the top of society—in the economy, in government, in education, in the criminal justice system, and in media—help perpetuate it. The institutions of society, the basic ways in which things are done, operate to uphold the existing inequality. The socialization of people into a culture that justifies inequality is also an important factor, as well as the successful socialization of people to learn and accept their position. Finally, instruments of force are used to perpetuate inequality.

Some inequality is probably inevitable. People must work very hard to prevent its emergence, and once it emerges, they must work even harder if they wish to control it. Robert Michels argues that organizations with leaders will inevitably develop a system of inequality and that ultimately that system will be very difficult to eradicate. Although Marx believed that with the destruction of capitalism equality would prevail in society, the twentieth century has not proved him right. Indeed, the situation seems far more complicated. In the Soviet Union, China, or Cuba, nations where private property was abolished, there still arose a stable system of inequality, perhaps not based on ownership of property as much as on political leadership, occupation, and control over (rather than ownership of) property. Even in the Israeli kibbutz, where people have equal control over the collective resources and where decisions are democratically made, an informal system of inequality develops between leaders and everyone else.

To claim that inequality is inevitable does not mean that people should also claim that poverty and hardship must be accepted or that tyranny must be tolerated. The question for all human beings should be, *How much inequality* is to be tolerated in society or in an organization? *How much* inequality is necessary? beneficial? democratic? humane? moral?

To realize that inequality is inevitable also means that those people dedicated to principles of equality have a difficult task ahead, because so much in society seems to encourage and protect a system of inequality. In this respect equality is like freedom: far from being automatic, it is possible only with eternal vigilance.

REFERENCES

The following works examine various forms of social inequality, focusing especially on class, race, and gender. All are good introductions to the general questions of why inequality arises and how it is perpetuated. Some look at inequality in all societies; some concentrate on the United States.

Adam, Barry B. 1978 *The Survival of Domination: Inferiorization and Everyday Life.* New York: Elsevier.

Baldwin, James 1963 *The Fire Next Time.* New York: Dial Press.

Ballantine, Jeanne H. 1989 *The Sociology of Education.* 2nd ed. Englewood Cliffs, NJ: Prentice-Hall.

Baltzell, E. Digby 1964 *The Protestant Establishment: Aristocracy and Caste in America.* New York: Vintage.

Beeghley, Leonard 1983 *Living Poorly in America.* New York: Praeger.

Beeghley, Leonard 1989 *The Structure of Social Stratification in the United States.* Boston: Allyn and Bacon.

Bernard, Jessie 1987 *The Female World from a Global Perspective.* Bloomington: Indiana University Press.

Blau, Peter M., and Marshall W. Meyer 1987 *Bureaucracy in Modern Society.* 3rd ed. New York: Random House.

Blauner, Robert 1972 *Racial Oppression in America.* New York: Harper and Row.

Bottomore, T. B. 1966 *Classes in Modern Society.* New York: Pantheon Books.

Bowles, Samuel, and Herbert Gintis 1976 *Schooling and Capitalist America.* New York: Basic Books.

Carmichael, Stokely, and Charles V. Hamilton 1967 *Black Power.* New York: Random House.

Cashmore, E. Ellis 1987 *The Logic of Racism.* London: Allen and Unwin.

Chambliss, William J. 1973 "The Saints and the Roughnecks." *Society,* *11*:24–31.

Chambliss, William J., and Robert Seidman 1982 *Law, Order, and Power.* 2nd ed. Reading, MA: Addison-Wesley.

Collins, Randall 1979 *The Credential Society: An Historical Sociology of Education and Stratification.* New York: Academic Press.

Cookson, P. W., and C. H. Persell 1985 *Preparing for Power: America's Elite Boarding Schools.* New York: Basic Books.

Dahrendorf, Ralf 1959 *Class and Class Conflict in Industrial Society.* Stanford, CA: Stanford University Press.

Dalphin, John 1987 *The Persistence of Social Inequality in America.* 2nd ed. Cambridge, MA: Schenkman.

Della Fave, L. Richard 1980 "The Meek Shall Not Inherit the Earth: Self-Evaluation and the Legitimacy of Stratification." *American Sociological Review,* 45:955–971.

Durkheim, Emile 1893 *The Division of Labor in Society.* 1964 ed. Trans. George Simpson. New York: Free Press.

Dye, Thomas R. 1986 *Who's Running America?* 4th ed. Englewood Cliffs, NJ: Prentice-Hall.

Eliade, Mircea 1954 *Cosmos and History.* New York: Harper and Row.

Engels, Friedrich 1884 "The Origin of the Family, Private Property, and the State." In *Karl Marx: On Society and Social Change.* 1972 ed. Ed. Neil J. Smelser. Chicago: University of Chicago Press.

Ewen, Stuart 1976 *Captains of Consciousness.* New York: McGraw-Hill.

Farley, John E. 1988 *Majority–Minority Relations.* 2nd ed. Englewood Cliffs, NJ: Prentice-Hall.

Feagin, Joe R., and Clairece Booher Feagin 1986 *Discrimination American Style: Institutional Racism and Sexism.* 2nd ed. Malabar, FL: Kreiger.

Freeman, Jo 1989 *Women: A Feminist Perspective.* 2nd ed. Palo Alto, CA: Mayfield.

Friedan, Betty 1963 *The Feminine Mystique.* New York: Norton.

Galbraith, John Kenneth 1979 *The New Industrial State.* 3rd ed. New York: New American Library.

Gilbert, Dennis, and Joseph A. Kahl 1987 *The American Class Structure.* 3rd ed. Chicago: Dorsey Press.

Gordon, Milton M. 1978 *Human Nature, Class, and Ethnicity.* New York: Oxford University Press.

Huber, Joan, and Glenna Spitze 1983 *Sex Stratification: Children, Housework, and Jobs.* New York: Academic Press.

Jencks, Christopher, et al. 1979 *Who Gets Ahead? The Determinants of Economic Success in America*. New York: Basic Books.

Kanter, Rosabeth 1977 *Men and Women of the Corporation*. New York: Basic Books.

Keller, Suzanne 1963 *Beyond the Ruling Class: Strategic Elites in Modern Society*. New York: Random House.

Kennickell, Arthur, and R. Louise Woodbum 1992 *Estimation of Household Net Worth Using Model-Based and Design-Based Weights: Evidence from the 1989 Survey of Consumer Finances*. Unpublished paper. Washington, DC: Board of Governors of the Federal Reserve System.

Kerbo, Harold R. 1983 *Social Stratification and Inequality: Class Conflict in the United States*. New York: McGraw-Hill.

Kitano, Harry H. L. 1985 *Race Relations*. 3rd ed. Englewood Cliffs, NJ: Prentice-Hall.

Lenski, Gerhard E. 1966 *Power and Privilege: A Theory of Social Stratification*. New York: McGraw-Hill.

Liebow, Elliot 1967 *Tally's Corner*. Boston: Little, Brown.

Lipset, Seymour Martin, Martin Trow, and James Coleman 1956 *Union Democracy: The Inside Politics of the International Typographical Union*. New York: Free Press.

Malcolm X and Alex Haley 1965 *The Autobiography of Malcolm X*. New York: Grove Press.

Marcuse, Herbert 1964 *One-Dimensional Man*. Boston: Beacon Press.

Marger, Martin N. 1987 *Elites and Masses*. 2nd ed. New York: Van Nostrand Reinhold.

Marx, Karl, and Friedrich Engels 1848 *The Communist Manifesto*. 1955 ed. New York: Appleton-Century-Crofts.

Matras, Judah 1984 *Social Inequality, Stratification, and Mobility*. 2nd ed. Englewood Cliffs, NJ: Prentice-Hall.

Michels, Robert 1915 *Political Parties*. 1962 ed. Trans. Eden Paul and Cedar Paul. New York: Free Press.

Mills, C. Wright 1956 *The Power Elite*. New York: Oxford University Press.

Mintz, Beth, and Michael Schwartz 1985 *The Power Structure of American Business*. Chicago: University of Chicago Press.

Myrdal, Gunnar 1944 *An American Dilemma*. New York: Harper and Row.

Noel, Donald 1968 "A Theory of the Origin of Ethnic Stratification." *Social Problems, 16*:157-172.

Olsen, Marvin E. (ed.) 1970 *Power in Societies*. New York: Macmillan.

Olsen, Marvin E. 1978 *The Process of Social Organization*. 2nd ed. New York: Holt, Rinehart and Winston.

Phillips, Kevin 1990 *The Politics of Rich and Poor*. New York: Random House.

Rousseau, Jean-Jacques 1755 "A Discourse on the Origin of Inequality." In *The Social Contract and Discourses*. 1913 ed. Trans. G.D.H. Cole. New York: E. P. Dutton.

Ryan, William 1976 *Blaming the Victim*. Rev. ed. New York: Vintage.

Ryan, William 1982 *Equality*. New York: Vintage.

Simpson, George E., and J. Milton Yinger 1985 *Racial and Cultural Minorities: An Analysis of Prejudice and Discrimination*. 5th ed. New York: Harper and Row.

Steinberg, Stephen 1989 *The Ethnic Myth: Race, Ethnicity, and Class in America*. 2nd ed. Boston: Beacon Press.

Tavris, Carol, and Carole Wade 1984 *The Longest War: Sex Differences in Perspective*. 2nd ed. New York: Harcourt Brace Jovanovich.

Thrasher, Frederic 1927 *The Gang*. Chicago: University of Chicago Press.

Tumin, Melvin M. 1985 *Social Stratification: The Forms and Functions of Inequality*. 2nd ed. Englewood Cliffs, NJ: Prentice-Hall.

Turner, Jonathan H., Royce Singleton, and David Musick 1984 *Oppression: A Socio-History of Black–White Relations in America*. Chicago: Nelson-Hall.

U.S. Bureau of the Census 1990. *Money Income and Poverty Status in the United States, 1989*. Current Population Reports. Series P-60, no. 168. Washington, DC: Government Printing Office.

U.S. Congress, Joint Economic Committee 1986 *The Concentration of Wealth in the U.S.* Washington, DC: U.S. Congress, Joint Economic Committee.

Useem, Michael 1984 *The Inner Circle*. New York: Oxford University Press.

Van den Berghe, Pierre 1978 *Race and Racism: A Comparative Perspective.* New York: John Wiley.

Wallerstein, Immanuel 1974 *The Modern World-System.* New York: Academic Press.

Whyte, William F. 1949 "The Social Structure of the Restaurant." *American Sociological Review, 54*:302–310.

Wilson, William J. 1973 *Power, Racism, and Privilege.* New York: Macmillan.

Wilson, William J. 1987 *The Truly Disadvantaged: The Inner City, the Underclass, and Public Policy.* Chicago: University of Chicago Press.

Wright, Erik Olin 1978 *Class, Crisis and the State.* London: NLB.

Zimbardo, Philip 1972 "Pathology of Imprisonment." *Society, 9*:4–8.

5

Why Do We Believe What We Do?

The Creation of Social Reality

*W*hy do I believe in God? Is it a belief that has been proved to me? Is it something that human beings believe naturally? Is it something that I have accepted from parents? Is it something that I find soothing? Is it something that I need to believe? Is it a truth that has been revealed to me from a supernatural source?

Try this question another way. If my life had been different, if I had been born at a different time or place, would I still believe in God? Would my beliefs about God be the same as they are now?

We believe a lot of things. For a moment consider what you believe. What do you believe about the death penalty? about human nature? about capitalism? individualism? freedom? about American high schools? about the frequent displays of violence on television? about men and women, romance, sex, and marriage? about the Middle East? about the future of the United States in the world?

Can you think of *any* idea you believe that does not have primarily a *social* foundation? Is there anything we believe that has not arisen primarily through *interaction* with others? This is the sociological approach to the question of truth.

The Utility of Knowledge

Psychologists and sociologists begin their approach to knowledge with the safe assumption that human beings have a limited capacity to remember things. Every day we are bombarded with facts and environmental stimuli. We take note of some things that matter at the moment; most things we do not notice at all. Of those things we take note of, we remember only a tiny number, and for most things our memory is very short. Students realize this when they study for exams: "I'll remember this for one day, but then I'll forget all of it as soon as I'm done taking the test!"

As we encounter new situations each day, we apply what we can remember, and we work through those situations using whatever seems to fit. Each time I walk into a classroom, I draw from all that I know and apply it to the topic at hand. When a question arises, I select from all my "facts" and give an answer. When I meet a political candidate at a party, I pull out what I think is relevant from my memory—always selective—and apply it in that situation. If I have learned to distrust political candidates, I ask questions to test the trustworthiness of this one. If I do not like Democrats and the candidate is a Democrat, I may see whether he or she is any different from other Democrats, or I may try to trip him or her up by asking questions that will reassure me and others that Democrats are still no good. On the other hand, perhaps the fact that the individual is handsome or young or recently divorced or Irish will be more important to me in the situation. I cannot possibly note everything, and I cannot possibly apply all that I have learned that might help me in that situation.

Out of all that I learn from others and from all the experiences I have had, I remember those things that are *useful* to me, those things that I can use in situations to achieve my goals and solve any problems that arise. And every time I apply something to a situation and it works for me, that success simply reinforces it, and I remember and use it again. No matter how truthful something is, I won't remember it if I do not use it. I believe in God because it is useful for me; I believe in a just God because that idea makes sense out of the world I encounter every day; I believe in a forgiving God because that helps me deal with issues of right and wrong; I believe in a loving God because that gives my life meaning. When these ideas no longer work for me, I forget them, and others take their place in my life.

The problem is to explain how ideas come to be useful to the individual, and so we enter the world of the *social*.

The Social Construction of Reality

Peter Berger and Thomas Luckmann, in their wonderful book *The Social Construction of Reality* (1966), show with great care how people come to create in interaction with one another the truths that they believe. They argue that groups of people find some ideas useful and reject or ignore ideas that are not useful. Fraternities teach the importance of sacrifice to the group in preference to individualism; charities teach sacrifice to the less fortunate; armies teach sacrifice to country rather than concern for one's own future. Individuals find these ideas useful largely *because* they are members of these groups. Belonging encourages them to see the wisdom of the teaching. The individual's useful ideas arise from ideas useful to the individual's groups.

The Meaning of Culture

Each organization—groups, formal organizations, communities, and societies—develops a set of ideas, values, and rules that come to be useful for achieving its goals and solving the problems with which it must deal. As we have seen, this is called *culture*. Berger and Luckmann point out that a hunting society will know facts about hunting, will develop a set of rituals around hunting, and will teach ideas concerning how to be a successful hunter. On the other hand, a community that is morally opposed to eating meat will know a different set of facts about hunting, will develop a system of morality that condemns hunting, and will teach ideas about how to survive without eating meat.

Culture contains our taken-for-granted truths, a set of assumptions that we generally accept without serious question. On a general level, a given culture may make religious assumptions or scientific ones about the universe. It may emphasize progress, or it may place great importance on tradition. It may be committed to the individual or to the collective. On this general level, culture tells us to value freedom, materialism, family, or art; it teaches us to work hard, take it easy, compete, cooperate, exploit others, or love them.

It is so general a guide, and so basic, that its elements seem natural to most of us, and it seems that something natural rather than social is at work.

Modernization is a good example of culture as a general guide to the way people think. When modernization occurs in society (that is, when industrialization, urbanization, and bureaucratization develop), a society's culture—and consciousness—is altered. People increasingly think of one another impersonally. They become more independent and anonymous. They value individuality and change. A people's view of time changes. They tend to think of the future more and pay less attention to the past. Instead of defining time circularly (tomorrow will be like the past; progress is an illusion), they define it linearly, on a time line with a beginning and a future (tomorrow will be different from the past; progress is possible). Even their definition of the clock changes. Instead of approximating days worked for purposes of pay, employers and workers place importance on hours and even minutes. Punctuality is stressed, as are discipline and routine. A people's culture is basic to their lives; it is a socially constructed set of ideas, values, and norms that forms the basis for the way they act.

Culture is best defined as a people's *shared perspective*. It is their approach to understanding reality. A culture is a *context* within which experience is perceived and interpreted. Of course, a shared perspective will be biased. It neglects so much, selects and emphasizes certain aspects of the environment, and always interprets facts in light of its preconceived assumptions. We cannot escape culture; we use it as a guide to construct our reality. And as we become committed to a culture, we have a strong tendency to see those who are different as outsiders, and if they threaten us, they become "deviant." We label them wrong, criminal, or sick. This designation, however, remains *social*; what becomes deviant is a result of our commitment to certain cultural standards.

Consider a gay couple, for example. Society teaches us about homosexuality, filling us with "facts," values, and morals that we use to react to the gay couple. Society gives us words—such as *gay, queer, faggot, lesbian*, and so on—that we use to apply to people who are homosexual. It teaches us reasons why people are gay (choice, weakness, illness, biology, upbringing, and so on), and it shows us why such actions are moral or immoral. Most of us are influenced by society's perspective, and so when we see a gay couple, this perspective becomes our guide for selecting what we see. It makes sense to us; it becomes useful.

Of course, not all of us think the same way, do we? If we are part of the gay community, we will learn a different perspective than if we are a member of a fundamentalist church. If we become part of a university community, chances are we will become more tolerant of human differences, and thus we will see homosexuality as simply sexual preference. If we are part of the community of psychiatrists, we will see another reality, and if we are sociologists, we will see yet another one. But that is exactly the point! We all interact—we take on the perspectives of those groups within which we interact—and those perspectives guide us in terms of what to believe about the world. *Reality never tells us exactly what it is; facts do not fill empty heads. Instead, human beings must interpret what exists, and that interpretation begins with the culture we use to guide us. That culture arises from our groups and from society.*

This view is not all that obvious to most people. It seems that we realize that other people teach us what we know, and thus we recognize the importance of social influence. But here we are looking at something much more basic. Whatever we learn is always interpreted *within a social context*—a culture—and that context strongly influences whether we believe what we learn, remember it, or use it. This idea might be illustrated in the following manner:

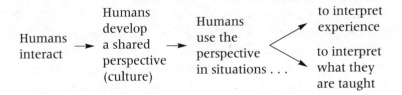

Remember: the perspective developed in a group is *useful* to the group; it *works*. If ideas, values, or rules somehow do not fit into what the group strives for, then they are forgotten or rejected. If we are a society of inequality, we develop ideas to justify that inequality or help us ignore it. If we are a capitalist society, it is important for us to value competition. If we have enemies, it is important for us to dehumanize them by calling them murderers, terrorists, or animals. If we are active in the right-to-life movement, it is important for us to believe that the fetus is a human being. Karl Mannheim writes that "even the categories in which our experiences are ... collected and ordered" depend on the position in society of the group whose ideas we use (1929:130). In every case a group's culture is linked to the group itself: what it does, how it is structured, what its activities are, and what its history has been. There is always a reason why a particular group believes in core cultural elements, apart from their truth or falsehood.

An example is fascism, a set of beliefs and values developed in Germany after World War I. Fascism is a political ideology, a set of ideas that exaggerates certain elements in the world in order to justify a certain political program—for example, a belief that war and power bring out the best in humankind, a belief that various subgroups of the population are naturally unequal, and a belief that democracy and freedom indicate weakness. Fascism became a central part of German culture because *it worked*: it explained the failure of Germany in the war, it explained the causes of the Depression, and it appealed to people's general discontent. It was consistent with many themes contained in traditional German culture: strong nationalism, militarism, and authoritarianism. It worked for certain categories of people such as German industrialists and various political opportunists who were able to use it to their advantage. Fascism worked because it explained the present, was consistent with the past, and gave hope for the future.

Perspectives are tricky. Once people believe one, they find it difficult to accept evidence that challenges it. An internal logic is at work, a selective interpretation of evidence, a tendency to interpret experience in line with what people already accept. It is not that people who disagree with us are fools; they simply have a different approach to the same world that we perceive, and they interpret events accordingly. It would seem that when someone predicts the end of the world and it does not come, followers would turn away, but there is evidence that followers become even more committed, because the faithful believe that their commitment saved the world. (See, for example, Leon Festinger, *When Prophecy Fails*, 1956.)

The concept that ideas simply "work" and that that is why we believe them is really too simple. Sometimes we continue to believe something we were taught earlier in life, or something that the media or political authorities want us to believe, and never wonder if something else might work for us much better. Until we fall into ideas that appear to us to work better than what we have been taught, we have a strong tendency to hang on to the old. Exposure to new groups and, therefore, new cultures presents us with such new ideas.

The Importance of Culture to the Individual

The social construction of reality, therefore, occurs because people in interaction with one another create a culture, or shared perspective, that works for them when they are together. But why does the individual come to take on that perspective? It is not inevitable, but the pressure is great. There are four reasons.

First, the individual will take on the perspective that works for him or her. As long as one is within a group, one will usually find its perspective useful: "People I know and with whom I interact every day use this perspective, and it works for them." The individual is encouraged and reminded to keep using the perspective of his or her group: "Its perspective is my perspective. Its concerns and values are mine, its rules and its answers work for me. Indeed, I stay in the group partly *because* its perspective works for me. The evidence is all around me. Those with whom I interact and whom I respect find it useful, and that influences me to believe that it's true."

Second, by its very nature, interaction works to reinforce the perspective that one learns from others. The more intense the interaction—that is, the more one interacts with certain others at the exclusion of outsiders—the more one will be convinced that the group's perspective is the most useful. After all, if I do not have the opportunity to test other perspectives, I have nothing to contrast the group's perspective with. Ongoing interaction reaffirms a perspective; it prevents one from going outside and learning an alternative perspective. An isolated community means fewer and more certain perspectives.

An urban, heterogeneous community usually means crisscrossing interaction, exposure to many groups, and less certainty concerning one's own perspective. In all cases, however, continuous interaction reinforces the perspective, and any breakdown in interaction undermines it. The significance of this connection can be profound when we recognize that interaction means *communication*, talking, writing, gesturing, and sharing language, ideas, values, norms, and so on. We develop culture within our communication channels. As they change, the culture that is important to us changes.

The third reason an individual takes on the group perspective is that the group encourages conformity through both continuous socialization and social sanctions. We are socialized to believe what the group does. We learn its perspective through listening and watching others and through intentional acts of teaching by others. Orientation to college life, basic training in the army, on-the-job training, and spring football practice are but a few examples of socialization into the culture of the group. We are also subject to social sanctions (rewards and punishments) from the group. When we use the group perspective, others in the group accept us, and we come to feel part of something important. If we do not use it, we are not rewarded, and sometimes we are punished. All social organizations want loyalty, and loyalty is tested through showing that we believe in the ideas and principles of the group.

Finally, we come to believe in the group's perspective because it is very difficult for human beings to believe something alone. We want to believe that our view of reality is *shared*. Most of us seek to check out our truths socially. Truth becomes socially anchored because we seek *social support* for our ideas. We are, in a sense, set up to take on the culture of those groups within which we exist; they help make us certain that what we believe about the world is true.

The first answer to why we believe what we do, therefore, has to do with *culture*. In interaction with one another, people develop and learn an approach (perspective, culture) for understanding the world. That approach is by definition selective and influences what we come to learn and how we interpret our experiences. We believe in our perspectives because we are active in the groups that share them. The groups that are useful to us share perspectives that become useful to us. Ongoing interaction, socialization, and reinforcement encourage us to believe in the culture, as does our need to have social support for our truths.

Social Structure and Reality

Positions and Perspectives

Our view of reality depends not only on the social organizations in which we are embedded but also on our *positions* in those organizations. Perspectives are points of view, angles from which one sees reality. One angle is from the top, another from the bottom. The upper class and the poor look at reality from different places, and these places are called positions. Leaders and followers in groups have different perspectives, and so do professors and students. The fact is that social organization creates what we believe not only through its culture (shared by everyone) but also through the positions that we take on and the roles that we play.

We believe, for example, in ideas arising from the fact that we are men or women. Part of how one becomes a woman in this society is to learn to "think like a woman," to believe, for example, that falling in love, marrying, and having children are necessary for a fulfilling life. To become a man is to learn to "think like a man," to believe, for example, that making money and gaining prestige, power, and privilege in the economic world are necessary for a fulfilling life. *Femininity* and *masculinity* are words associated with behaviors expected of people, and along with those behaviors go ways of thinking. In our more open society today, of course, this distinction is no longer simple, but the increasing complexity does not contradict its existence. It is easier to see these roles in a religious fundamentalist culture (Muslim, Jewish, or Christian). It is probably inconceivable to most of us how traditional Iranian women can allow themselves to be subservient to men, to reject utterly doing the kinds of things that men do, to be satisfied with being solely wives and mothers, and even to cover their faces and bodies when they go out in public. If we were to ask them, however, we would find a belief system that explains it. And once we understood that belief system, we would understand the logic of such actions. Women are not simply forced to do such things; they are taught a perspective that justifies it. And their angle of vision becomes part of their very being; they come to think differently from the men with whom they interact. This would also be obvious to us if we examined Jewish or Christian fundamentalism. Let us not forget that in any society women and men have different positions (they are expected to perform different roles and are granted different amounts of power, privilege, and prestige in society) and therefore will have different perspectives. The perspective of women will influence individual women's ideas concerning pregnancy, abortion, marriage, birth control, menstruation, occupational opportunities, American history, and professional athletics.

But go beyond gender. A factory worker thinks differently from management; a boss, from an employee; an owner, from a manager; a bookkeeper, from a secretary. Each has a position in the social structure, and each has a different way of looking at reality. Imagine society as having thousands of positions, occupational and otherwise. We interact and find ourselves in these positions. And what happens to us? We come to think of the world according to these positions: high school graduate, dentist, artist, general in the army, ex-prisoner, member of the upper class, African American, rock star. They influence how we look at reality: at the world out there as well as our own inner world. It is difficult to be in a position, play the role that is expected, yet escape the tyranny of the perspective that goes with that position. The suddenly successful rock star matter-of-factly declares: "I'm just the same person I always was; I think the way I always did." This could be true, but usually it is not. Peter Berger (1963) reminds us that it is difficult to lie in this world. We come to believe in those roles that we play.

Why do we do it? Why must we necessarily take on the perspective of the positions that we occupy? The simplest answer has to do with the explanation I have used throughout this chapter: the perspective associated with a position is *useful* for those in that position. It works. To play a role associated with a position we must know how to think in that role—sometimes even when we do not like the role. If I am going to succeed as a student, I must take on the perspective of a student and understand the world from that angle. If I am going to be a boss or a manager or a quarterback or a president, I had better think like one. If I go to prison, I had better know how prisoners think, or getting through that ordeal will be much more difficult. As our positions change throughout our lives, our perspectives must change, or we will not successfully play the roles associated with those positions.

We are socialized to think according to our positions. If we become a prisoner, other prisoners interact with us, and we learn how they think. The guards and warden interact with us, and they, too, tell us how to think. We come to learn how "people like us" are supposed to think, and we are influenced. Socialization strikes again!

But there is another subtle and extremely important reason why positions are so powerful in influencing our thought. Each position sets up the occupant to see reality from that particular angle. In a sense, a student must see the world from a student's angle; that is the place he or she is looking from. If I am a man, I cannot easily think like a woman. I can try to understand how women see the world, but that is not the position I am in, so I do not regularly look at the world from that position. I cannot easily think like a poor person or a rich person unless I am one. I cannot know what it means to be nonwhite in American society, a general in the army, or the president of the university if I am not in these positions. All my efforts at studying other positions do not quite cause me to think like the people in them, for when you are *in* a position and you see the world from that position, you have an angle on reality that no one else can really have.

Human beings change their perspectives when they occupy new positions, when they go from student to graduate, from single to married, from working class to middle class. We can try to fight changing our perspectives as we change positions ("Aw heck, I'm just the same guy I used to be"), but we usually change without even realizing it. Positions bring us perspectives, we find them useful, others socialize us into these positions, and, probably most important, positions bring us angles on reality that are difficult to escape.

Age Cohort as a Position in Society

Let us expand the meaning of position to include *age cohort*, the individual's generational position in society. The baby boom generation is that category of people, born right after World War II, that reached its forties in the 1980s. This generation is in a different societal position, and that forms its perspective.

When you were born matters, for it places you within a historical period when a generation begins to form its perspective, it ties you to that generation's perspective throughout your life, and it influences much of what you believe. Those who experience war in their generation have a different view of war (and peace) than those who do not. Those who had job opportunities when they graduated from college have a different view of work and the future than those who did not.

We learn our views of family and sex at a certain point in our society's history. Divorce, birth control, abortion, marriage, and gender equality will have different meanings for people depending on which generation they are part of: "I can't help it. When I grew up, that's what I learned about sex!"

Those who were young in the 1950s and '60s remember the civil rights movement. It was part of their experience, and they were influenced by events reported daily in their newspapers. They saw what people were fighting against, and they interacted and shared a perspective that they used to interpret that fight. It is now hard for them to explain to generations who grew up afterward what the fight was all about.

Think for one moment how different your perspective would be if your generation had fought in World War II, Korea, or Vietnam; if your generation had been workers fighting for unionization in the 1920s and '30s; or if it had been the one that emerged out of World War II ready to reap the benefits of an affluent and highly conformist society. Or think how different one's perspective is depending on when his or her generation went to college: in the '50s, when teaching and engineering were highly valued professions; in the '60s and '70s, when social work, law, and medicine were emphasized; or in the late '70s and '80s, which stressed computers and business. In many basic ways our view of what a successful life entails is influenced by the perspective that emerges in our particular generation.

Power and the Creation of Perspectives

Throughout this chapter I have emphasized two central points: people take on the culture, or shared perspective, of their group, and people take on the perspective associated with their positions. The perspectives of their groups and their positions shape their reality. These perspectives cause people to believe selectively certain things and to perceive selectively a world both out there and within themselves.

What I have not emphasized is that perspectives are often shaped to some extent by the powerful in society. Marx and Mannheim note how the powerful create ideologies, exaggerated and even outmoded perspectives on reality used for the purpose of defending the status quo, that is, their own power, privilege, and prestige in society. Dictators teach people to obey, arguing that order is necessary now and that obedience will someday bring prosperity for all. The upper class teaches people that what it has it deserves and has worked hard for. Slave owners teach slaves and others the justification for slavery, especially a religious justification, and racists inevitably have a whole set of ideas they try to teach others to justify inequality. The ideas are often sophisticated, and even those who are poor or slaves or victims of racism may come to believe them.

The perspective of the poor, for example, is usually a complex mixture of beliefs developed in their position as they interact and live that position. But it is also the result of ideas developed by wealthier individuals with whom they come into contact directly (in interaction) or indirectly (through media). The poor are often conservative in society, even though they have little to gain from the way society is. Because it is not the poor who control advertising and educational or political institutions, it is not their ideas that are taught in society, so they, like everyone else, become influenced by the ideas of those who do control these avenues of influence.

This principle works throughout society. Those who have power in any organization will influence how others think. The students in any university may develop ideas in interaction with one another, but they are also socialized to believe the ideas (about what students are to believe) from faculty members and administrators. Americans are influenced by the advertising of large corporations and the speeches of their political leaders. More than this, our perspective as consumers is molded by the nature of all advertising: buy, charge, have a better life through better products, and always seek the new, the beautiful, and the exciting. What we are to think as parents, men, women, young people, lovers, friends, thirtysomething yuppies, or stars on a basketball team is influenced by a complex mixture of advertising, situation comedies, soap operas, music, and college textbooks. Much of this thinking results from powerful economic and political actors whose interests are served by our accepting these ideas. The perspectives we use, therefore, must always be understood, in part, within a larger context of power. Through a complex mixture of coercion and manipulation, powerful people in every organization mold—at least to some extent—the culture of the organization as well as the ideas associated with the various positions in it.

The Changing Nature of the Individual's Reality

Where does this discussion leave us? Why do we believe what we do?

We believe what is useful to us. The usefulness of knowledge is a result of the perspectives, or cultures, we share in interaction. It is also a result of perspectives that arise from the positions we hold within social organization.

But we all change, don't we? Why is it that we believe some things at one time in our life and are able to overcome those beliefs and take on new ones? This is, of course, a difficult question, but it, too, makes sense within a social context.

We change what we believe because we interact with new people and enter new organizations. Belief is embedded within our social life, and our social life is continuous. We always interact, and we join new groups, formal organizations, communities, and even societies. The human being is an actor on many stages, each one bringing a slightly new perspective and thus calling for a slightly different set of truths to interpret reality. As we move from one stage to another, we change. The culture of the elementary school is different from the one we take on in high school, and both are different from the one we take on in the family or at our first job.

We also change because culture changes; in interaction it is constantly being questioned, challenged, and altered, usually a little but sometimes a great deal. This is especially true in small groups, but it is certainly true of society's general culture, too. Witness the rapidity with which the former Soviet culture is changing today. Witness also how our ideas of family life, gender inequality, and the socialization of children have changed in the past twenty years. Fifteen years ago the term *child abuse* was not even a part of our normal vocabulary; some people thought that beating children was part of the way to teach discipline. Today even spanking children is suspect.

We also change because we change positions. We believe things as seniors that we did not believe as freshmen. We believe things as president that we did not as vice president. We believe things as a married person that we did not believe as a single person. We change a lot, as new positions demand new perspectives for interpreting reality.

Indeed, we change simply because we grow older. Age is a position in society. Childhood, preadolescence, adolescence, young adulthood, middle age, and old age are all stages in people's lives that bring on their own perspectives and interpretations of reality. Our view of retirement changes as we grow older, as does our view of the meaning of life. Religious commitment changes as we age, and our political ideas change. Our view of the draft depends on how old we are, as does our view of AIDS.

Belief, therefore, is never fixed. It changes as we interact and change organizations; it changes as culture changes; it changes as

we change positions. Change occurs slowly, without our even realizing it. It is difficult for us to wake up one day, look back, and ask
in earnest, "How could I ever have believed that?"

The Importance of Our Past

We change, yet it is also impossible to escape our past. Human beings interact throughout their life, constantly altering perspectives
and ideas, but the past is never erased.

Primary socialization is the socialization of small children in
their home and neighborhood. There we first learn about the world,
take on our first perspectives and ideas, and develop our first rules
for self-control. Secondary socialization comes from our school, the
media, our friends, and anyone else who tries to shape who we are
and what we think after this early primary socialization. Secondary
socialization normally does not radically change who we are, because
primary socialization normally remains centrally important to our
belief system. That is because it is *first*. We learn language, a system
of categorizing the world (men, women, boys, girls), a set of ideas,
and our first perspectives. We have nothing to compare these early
views with. We are not in a position to judge their worth. Continuous interaction with very powerful people (parents), whose affection
most of us seek, reaffirms the truth of these ideas, and, at least for a
while, we believe the ideas because we see no alternatives.

When alternatives do come in secondary socialization, a perspective and beliefs are already there, and the individual must compare and contrast new things with what he or she already knows.
The new is sometimes accepted, but not quite as unquestioningly as
the old. The old hangs on until it is replaced by something that seems
better. The old becomes a filter through which new truths are
judged, accepted, or rejected, and thus the old exists as a powerful
guide to what we believe. The earlier our perspectives and their
truths, the more difficult it is to erase and replace them. The burden
is always on the *new*.

Each group we enter, however, has something to do with what
we believe. To some extent, belief is situational, useful for the moment, useful in the group we happen to be in. Groups exercise pow-

erful forces on the individual to conform to the dominant view. Studies by Schachter (1951) clearly show that it is those who conform to the dominant belief of the group who are liked the best. Studies by Asch (1951) show how people change their expressed views, probably because they do not wish to be embarrassed by being different. Groups influence perceptions, ideas, values, emotions, and decisions. Irving Janis (1982) shows us how groups tend to move toward a consensus, discouraging disagreement and often making poor decisions because people are afraid to disagree with what they think their leaders want.

Thus, all of us are complex mixtures of past and present, of beliefs learned early through primary socialization and beliefs that conform to the groups we happen to be in at the moment. We change as new groups give us perspectives that cause us to question old beliefs, but change is often difficult. In fact, we should recognize that to a great extent the present simply reinforces the past; secondary socialization reaffirms much of what we learned in primary socialization; the culture of early organization prepares us well for later organization.

Do Individuals Form Their Own Ideas?

By now it must be obvious that the reasons we believe what we do are complex and manifold. All have in common a central principle of sociology: reality is socially constructed. We see the world, we interpret experience, and we learn ideas through perspectives that we take on in interaction with others. The cultures from which we learn and the perspectives associated with our positions (from social class to position in formal organizations to age cohort to age itself) shape what we see in the world. We change, but change should be placed in the larger context of our social life. We change, but our earlier primary socialization helps ensure stability of belief as we come in contact with new groups and new perspectives.

Our discussion now becomes even more complex. Are our beliefs simply the result of our social life, or do we have any control over them? The case thus far is a strong one: when we understand the influence of perspective over what we believe and see how closely tied it is to our social life, it becomes difficult to see what could cause us to develop our own ideas.

However, sociology is a complex perspective. Although some sociologists see human thought as simply the product of society, most see human beings as capable, at least to some extent, of developing their own perspective and ideas.

To understand this capability, we might return to the argument that began this chapter. People believe that which is *useful* to them. Usefulness is tied to our social life, but it is also tied to a testing that we all do as we deal with our life situations. Our actions test out what we know and believe. If our perspectives and beliefs help us achieve our goals, the test is positive, and we continue to believe. If they do not work, however, we begin to question their truthfulness. Human beings, although influenced by their associates, are ultimately their own testers. We check out what we are taught, and if we alter what we know, we can eventually end up openly questioning what others teach or can leave the groups whose perspectives simply do not work for us.

Furthermore, although the power of our social life is never to be denied, there is also the particular human quality of *thinking and interpretation*. Human beings live in a world that they give meaning to, define, analyze, and wonder about. This activity means that whatever we are taught is not necessarily accepted but can be—and often is—questioned, altered, integrated into what we know, or rejected altogether. We are in part responsible for our own ideas, as strong as our social life is, because we also have this active side.

Summary and Conclusion

Why, then, do we believe what we do? We are socialized, we take on the perspectives of our society and groups, and we see the world from the positions we take on. Our age matters, our age cohort does, and the power of primary and secondary socialization is not to be denied. Yet we also think and wonder about what we are taught, we apply it and test it out, and this process allows us to develop our own ideas, at least to some extent.

References

The following works examine the formation of ideas and culture in society. All of these works focus on socialization.

Aronson, Elliot 1988 *The Social Animal.* 5th ed. San Francisco: W. H. Freeman.

Asch, Solomon E. 1951 "Effects of Group Pressure Upon the Modification and Distortion of Judgments." In *Groups, Leadership, and Men.* Ed. Harold Guetzdow. New York: Carnegie.

Baldwin, James 1963 *The Fire Next Time.* New York: Dial Press.

Ballantine, Jeanne H. 1989 *The Sociology of Education.* 2nd ed. Englewood Cliffs, NJ: Prentice-Hall.

Ball-Rokeach, Sandra J., Milton Rokeach, and Joel W. Grube 1984 *The Great American Values Test: Influencing Behavior and Belief Through Television.* New York: Free Press.

Becker, Howard S. 1953 "Becoming a Marihuana User." *American Journal of Sociology,* 59:235–247.

Berger, Peter 1963 *Invitation to Sociology.* New York: Doubleday.

Berger, Peter, Brigitte Berger, and Hansfried Kellner 1974 *The Homeless Mind: Modernization and Consciousness.* New York: Vintage Books.

Berger, Peter L., and Thomas Luckmann 1966 *The Social Construction of Reality.* New York: Doubleday.

Blumer, Herbert 1969 *Symbolic Interactionism: Perspective and Method.* Englewood Cliffs, NJ: Prentice-Hall.

Brim, Orville G., Jr. 1968 "Adult Socialization." In *Socialization and Society.* Ed. John A. Clausen. Boston: Little, Brown.

Brim, Orville G., and S. Wheeler (ed.) 1966 *Socialization After Childhood.* New York: John Wiley.

Charon, Joel M. 1995 *Symbolic Interactionism: An Introduction, an Interpretation, an Integration.* 5th ed. Englewood Cliffs, NJ: Prentice-Hall.

Clausen, John A. 1986 *The Life Course: A Sociological Perspective.* Englewood Cliffs, NJ: Prentice-Hall.

Durkheim, Emile 1895 *The Rules of the Sociological Method.* 1964 ed. Trans. Sarah A. Solovay and John H. Mueller. New York: Free Press.

Durkheim, Emile 1915 *The Elementary Forms of Religious Life.* 1954 ed. Trans. Joseph Swain. New York: Free Press.

Eliade, Mircea 1954 *Cosmos and History.* New York: Harper and Row.

Elkin, Frederick, and Gerald Handel 1984 *The Child in Society: The Process of Socialization.* 4th ed. New York: Random House.

Ewen, Stuart 1976 *Captains of Consciousness.* New York: McGraw-Hill.

Feagin, Joe R. 1975 *Subordinating the Poor: Welfare and American Beliefs.* Englewood Cliffs, NJ: Prentice-Hall.

Festinger, Leon 1954 "A Theory of Social Comparison Processes." *Human Relations,* 7:117–140.

Festinger, Leon 1956 *When Prophecy Fails.* Minneapolis: University of Minnesota Press.

Fine, Gary Alan 1987 *With the Boys: Little League Baseball and Preadolescent Culture.* Chicago: University of Chicago Press.

Geertz, Clifford 1965 "The Impact of the Concept of Culture on the Concept of Man." In *New Views of the Nature of Man.* Ed. John R. Platt. Chicago: University of Chicago Press.

Hertzler, Joyce O. 1965 *A Sociology of Language.* New York: Random House.

Hostetler, John A. 1980 *Amish Society.* Baltimore: Johns Hopkins University Press.

Janis, Irving L. 1982 *Groupthink.* 2nd ed. Boston: Houghton Mifflin.

Jones, Landon Y. 1980 *Great Expectations: America and the Baby Boom Generation.* New York: Coward, McCann and Geoghegan.

Jones, Ron 1981 *No Substitute for Madness.* Covelo, CA: Island Press.

Liebow, Elliot 1967 *Tally's Corner.* Boston: Little, Brown.

Lofland, John 1966 *Doomsday Cult.* Englewood Cliffs, NJ: Prentice-Hall.

Lynd, Robert, and Helen Lynd 1929 *Middletown.* New York: Harcourt Brace Jovanovich.

Lynd, Robert, and Helen Lynd 1937 *Middletown in Transition.* New York: Harcourt Brace Jovanovich.

Malcolm X and Alex Haley 1965 *The Autobiography of Malcolm X.* New York: Grove Press.

Mannheim, Karl 1929 *Ideology and Utopia.* 1936 ed. New York: Harcourt Brace Jovanovich.

Marx, Karl 1848 *The Communist Manifesto.* 1955 ed. New York: Appleton-Century-Crofts.

McCall, George J., and J. L. Simmons 1978 *Identities and Interactions.* New York: Free Press.

Mead, George Herbert 1925 "The Genesis of the Self and Social Control." *International Journal of Ethics, 35*:251–277.

Mead, George Herbert 1934 *Mind, Self and Society.* Chicago: University of Chicago Press.

Rose, Peter I. (Ed.) 1979 *Socialization and the Life Cycle.* New York: St. Martin's Press.

Roszak, Theodore 1969 *The Making of a Counter-Culture: Reflections on the Technocratic Society and Its Youthful Opposition.* New York: Doubleday.

Ryan, William 1976 *Blaming the Victim.* Rev. ed. New York: Vintage.

Schachter, Stanley 1951 "Deviation, Rejection, and Communication." *Journal of Abnormal and Social Psychology, 46*:229–238.

Shibutani, Tamotsu 1955 "Reference Groups as Perspectives." *American Journal of Sociology, 60*:562–569.

Shibutani, Tamotsu 1961 *Society and Personality: An Interactionist Approach to Social Psychology.* Englewood Cliffs, NJ: Prentice-Hall.

Shibutani, Tamotsu 1986 *Social Processes: An Introduction to Sociology.* Berkeley: University of California Press.

Sykes, Gresham M., and Sheldon L. Messinger 1960 "The Inmate Social System." *Theoretical Studies in Social Organization of the Prison.* Pamphlet 15. New York: Social Science Research Council.

Weber, Max 1905 *The Protestant Ethic and the Spirit of Capitalism.* 1958 ed. Trans. and ed. Talcott Parsons. New York: Scribner's.

White, Leslie A. 1940 *The Science of Culture.* New York: Farrar, Straus and Giroux.

Whyte, William Foote 1955 *Street Corner Society.* Chicago: University of Chicago Press.

Williams, Robin 1970 *American Society: A Sociological Interpretation.* New York: Alfred A. Knopf.

Yinger, Milton J. 1982 *Countercultures: The Promise and Peril of a World Turned Upside Down.* New York: Free Press.

6

Are Human Beings Free?

The Possibility for Freedom in Society

*F*or much of my life I have pondered the question of freedom. In adolescence I asked myself if one could believe in God's will and in freedom simultaneously. As I studied the history of the world, I became aware of the human struggle for freedom, yet it bothered me to learn that psychology and sociology seriously questioned the possibility for much freedom. Sigmund Freud, Karl Marx, and Erich Fromm revealed to me new dimensions to the problem, challenging my understanding of what the concept of freedom means, and even our desire to be free.

From all of my studying and thinking, my most important conclusion is that it is impossible to determine whether human beings are free. This conclusion is difficult for me to accept, but it is, in fact, what I have learned. The study of sociology has had a great deal to do with this conclusion, and it has influenced most of my other beliefs about freedom. I am relatively certain about three other ideas:

1. If freedom exists, it is always limited by social forces that most people are only barely aware of.

2. Most people have a highly exaggerated view of how much freedom they actually have.

3. It is important in U.S. society for people to believe that Americans are free and that most others are not.

The study of sociology will cause many students to think critically about their beliefs about freedom. Sociology is a social science, and its goal must be to try to understand *why things happen*, why human beings act as they do. To ask the question of *why* is to look for forces at work on the individual—in sociology those forces are social—and to understand action in the context of those forces. An intelligent analysis of cause will include the question: To what extent do people actually control their own choices?

The Meaning of Freedom and Responsibility

To understand to what extent sociologists believe that human beings are free and responsible for their actions, we must tackle the difficult topic of what sociologists mean by the concepts of freedom and responsibility.

What happens when someone does something we do not like? We get angry, and we blame the individual. This tendency to blame others for what they do simply means we believe that they somehow had control over their acts and acted deliberately: "He knew what he was doing!" "She knew that others would be hurt. She doesn't care." "It was his fault that he got her pregnant." "No, it was her own fault." In a sense, we justify our own anger toward what others do by assuming that they had control over their actions. The judges at the Nuremburg Trials after World War II ruled that human beings had acted immorally and that the excuse that they had only been following orders was unacceptable. Instead, they were found guilty of *choosing* to do evil things against humanity: "They knew what they were doing, and they could have said no."

The court system in any society is generally based on the principle that human beings are *in control* of their actions. Pleading insanity means that they are deemed not in control and therefore not responsible for their actions. Sending people to prison or executing them for transgressions usually assumes freedom. Western religion assumes that punishment and reward await us after death primarily because of the choices we make during life, and this assumption implies that we control our own lives.

The concepts of freedom and responsibility both have to do with control over one's own life. Having *freedom* means that individuals control what they think and do. To say that individuals are *responsible* for what they do means essentially the same thing: they are in control. What is meant by control? To control one's life means that one understands the choices one makes, understands the options and consequences, and is in the position of actively determining those choices. To be free, therefore, means knowing choice, making choice, and ultimately controlling one's directions according to choice. If this characterizes what human beings are, we can say that they are free, or are responsible for their acts. And when we declare that people should *take responsibility* for their actions, we mean that they should realize that they are in control and should face the fact that their actions resulted from their free choice.

Sometimes we mean something else by *responsible*. It is often said that "humans should be free, but they also should be responsible." This statement means something different from control over one's life. It means that when I act, my choices should not be completely selfish, my actions should conform to a social morality—in short, that I should act according to conscience. In this sense, believing that the human being should be free is accompanied by the belief that freedom must always be limited by the individual's conformity to social considerations.

To act freely, therefore, means that individuals control their own thoughts and actions. To be *responsible* for action means that they are free. To *take responsibility* for their actions means that they recognize and accept the fact that they are free. To say that people should *act responsibly* does not mean that they should be free; it means that when they act, they should conform to a body of rules: freedom is all right as long as free acts are performed within a certain moral context.

Freedom as a Value in American Culture

There is little question that people in the United States believe that they are free and that it is very important to be free. But believing that one is free does not make one free. Indeed, it is important to realize that believing that one is free may stand in the way of being free, may actually prevent one from controlling one's own life. For if we believe that we are already free, efforts to alter whatever stands in the way of achieving control over our own lives seem unnecessary.

Freedom seems real to most Americans. Our wars are fought in the name of freedom, immigrants come to the United States in search of freedom, government justifies much of what it does by claiming that it is defending our freedom, and we generally believe that those who are successful and those who are not get their just rewards because they have freely chosen what to do with their lives.

It is very important for people in the United States to *feel* that they are free. Indeed, it is important for most people to feel this way. It means that they control their own destinies. If we did not believe these things, much of what happens in society would not make much sense to us: "After all, if we're not really free, what is the difference between our society and the rest of the world?" "What are the revolutions in the Soviet Union and Eastern Europe all about?" "If people are not really responsible for what they do, how can we morally justify capital punishment, or even most other instances of punishment (except as a means of rehabilitation or as an expression of simple revenge)?" "If we are not free and responsible for our own actions, does salvation make any sense, does life have any real meaning, or do we have any right to take pride in what we do?" "If I am not responsible for my own acts, then I really shouldn't care about my life—after all, it actually isn't *my* life."

It is important for every society to hold people responsible for their actions, assuming that they know what they are doing and punishing those who make evil choices. Society works only because such ideas exist and are an integral part of *culture*. Freedom is a central value in our culture, and we have a great many ideas that defend its existence. However, our question remains: Are human beings really free?

The Sociological View: The Power of Society

The Sociological Dilemma

Sociologists are caught in a great dilemma: they want to believe that human beings are free, yet they understand so well how all-powerful society seems to be. Sociologists like to claim that "society shapes the individual, but the individual shapes society." But when it really comes down to it, their work shows all the ways in which society shapes us and very few ways in which the individual shapes society. Emile Durkheim was a champion of freedom within society, yet his work shows us how powerful society is and tells us almost nothing about how freedom is really possible—neither how the individual can learn real control over his or her own life nor how such real control can even be allowed in society.

Peter Berger shows us that the individual is subject to social controls, social stratification, social institutions, socialization, roles, and groups, yet he claims in his writings that one purpose of sociology is to "liberate" human beings through helping them understand all the controls that shape them. One is left, however, with a basic question: How much freedom is possible within this all-powerful prison of society? Is any significant freedom possible? Marx, too, yearns for freedom and liberation for all humankind, yet almost all of his work shows the power of society over the individual. He maintains that someday the workers will control their own lives, after a revolution and the overthrow of capitalism. Yet studying Marx normally leaves one with the recognition of the enormous power of society rather than with much hope of liberation.

Social Problems, Social Rates, and Freedom

Perhaps we might continue the sociological approach to freedom by considering an argument made by C. Wright Mills in *The Sociological Imagination* (1959). Mills argues that to think sociologically is to see oneself *located* in both society and history. It is to understand that one exists within a social context that has developed over many years. One may experience personal problems—a bad marriage, overwhelming debt, unemployment, a midlife crisis—but such problems must be perceived in the larger context of society if one is to understand why they occur. The personal problems I experience are not mine alone. I experience the same problems as many other people in my position. If I had lived in another age, my problems would have been different. If I lived in another social position in this society, my problems would also be different. The social problems of my society at this particular time set me and others like me up for my personal problems. Although everyone in my position will not have my problems, the nature of society today makes these particular problems much more probable than they would have been in the past (or in another society). The particular problems that individual African Americans face in their personal lives exist because racial hatred, racial discrimination, and lack of decent opportunities exist in the society within which they were born. To live in a society in which child abuse or spouse abuse is common and has historically even been legitimated means that one lives within social forces that encourage many of us to be abusers or to be the victims of abusers.

Personal problems are linked to the nature of society itself. For example, people do not simply get divorced or commit suicide or commit crime in a random manner. If this were so, each year we would have a different number of such individual events. In fact, such problems occur according to fairly predictable and stable rates. We know that about half of all marriages today will eventually end up in divorce, because forces in society produce such rates. We know that approximately 19 males out of 100,000 will commit suicide, because the various forces at work in society produce about that many suicides. We have a birthrate, death rate, rate of migration, unemployment rate, school-dropout rate, crime rate, and pregnancy rate. In each case it is clear that society is a powerful force influencing the decisions of the individual. Take any society, any community, any neighborhood: identify the various rates, and we can better understand why people make the choices they do there. The individual who fools the social forces does not disprove their existence. Usually, it can be shown what social forces are necessary for the individual to rise above those that control so many others.

The existence of *social problems* and *social rates* is one starting point for sociologists in unraveling the question of freedom and responsibility. Where and when we are born subject us to certain social problems and social rates and influence the directions we take in life. If I am born in a ghetto, where the teenage pregnancy rate is extremely high, the chances are higher that I, too, will become pregnant as a teenager; if I am born outside that community, the chances are higher that I will not. Freedom? Maybe, but we do not choose the communities within which we are born, and if we are subject to such rates and problems, we have to work extra hard to go in a direction that others take for granted.

Poverty is an example of how the individual's destiny is influenced by social problems and rates. Although it is common for the general public to argue that poverty exists because people freely choose that direction, sociologists rarely make this claim. Such thinking is called "blaming the victim" for a serious social problem. Who becomes poor? Many people are born into poverty. Many children are poor: 20.7 percent of those under the age of eighteen in 1985; 40 percent of all Hispanic children; 44 percent of all African-American children (Jacobs, Segal, and Foster, 1988:71–72). Many poor people are women who are single parents, victims of desertion or divorce (about half of all poor families are households headed by women with no husband present). Many are elderly. Many are minorities (31.1 percent of all African Americans are poor, as are 29 percent of all Hispanic Americans). Many are people put out of work because of the closing of places of employment. Many are people in towns and farms left behind by rapid social change.

Is the poverty that comes to these people a free choice made by them? Do people brought up in a community where public education fails the vast majority of people freely choose to drop out of school? Do women brought up in a community where so many other teenage women get pregnant freely choose to get pregnant? Do people choose to be laid off or freely choose to live in communities where a business moves out because it cannot make it? Do people freely choose to live in a society where their racial group has minority status? Indeed, if I grow up farming, do I freely choose to become a farmer, destined to be put out of business because I cannot compete with large corporate farms?

The public, however, tends to see poverty as resulting from free will. For example, Joe R. Feagin (1975) studied American beliefs about the causes of poverty and found, not surprisingly, an emphasis on individual will rather than on social cause. By far, people regarded reasons such as "poor money management," "lack of effort, talent, or ability," or "loose morals and alcoholism" as good explanations of poverty (more than 80 percent in each case). Bad luck, being taken advantage of, failure of private industry, discrimination, and poor schooling were significantly less important (35 to 60 percent in each case). Clearly, the more personal the reason, the higher it was rated.

Our whole lives are influenced by many social factors: other people around us, socialization, our social class, our social groups, and social institutions, for example.

Take the most personal of our decisions: marriage. We claim the right to make a free choice. But how free is it, given the many social influences on our choice? Indeed, how free are we in our decision to marry in the first place? Bert Adams (1979), a sociologist, put together a theory that summarizes research studies and tries to explain the factors influencing whom we marry. Choice of mate is influenced by all of the following factors: physical proximity (we marry someone who happens to be within one of our social worlds), reaction by significant others to the relationship, similarity in physical appearance, similarity in personality, homogeneity in background (class, race, ethnic group, and religion, for example), absence of unfavorable parental intrusion, lack of alternative attractions, compatibility, and the perception of others that the two people really constitute a pair. Adams lists more personal factors, too (such as physical attractiveness and similar interests in early stages of the relationship), but here I am focusing primarily on social factors, which are also the focus of his theory. We *may* have some free choice, but social factors narrow our choice considerably.

Freedom: Thought and Action

To repeat: Freedom has to do with controlling one's own life. It means that other people do not control one's life. It means that society and social organization do not control one's life.

For purposes of understanding, it is important to divide freedom into two parts. First, if one is free, one is free in one's *thinking*. The ideas, values, and rules I believe in are my own—not society's, not other people's. The decisions I make in my head are my decisions; the truths I believe in are my truths. Without freedom to think, freedom to act is an empty freedom, because it means acting according to what others believe and have taught me to believe. If, for example, I claim the freedom to speak or write but the thinking that goes into what I say or write is not my own, the freedom to speak or write is meaningless. If I claim the freedom to make money but all around me I am influenced by others' desires to want to make money, then what real freedom do I have? Others have told me what to do with my life.

The second aspect of freedom involves *action*. I may be free to think but not free to act according to what I think. I may oppose my government, but unless I can *do* something to protest its policies, my freedom is not meaningful. I may decide that clothes are not right, but unless I am free to go naked, my freedom is not very important. I may believe that a university education is important, but unless I have the abilities and the finances necessary, I am not able to go to a university. I may decide to be a doctor, a lawyer, a teacher, or a musician, but my actually becoming one of these depends on much more than thinking: it depends on whether I am able to act effectively on my thinking. I may believe that giving my children many opportunities is important, but if I do not have the financial means to do so, my belief is not easily translated into action.

Although the individual may have some freedom to think and act, the typical sociological approach is to show the many ways in which freedom is curtailed. Few sociologists claim that there is no freedom; but all point out the many ways in which freedom is limited by society through socialization, social patterns, and various institutions of social control. In the next two sections we will examine two types of social control, over thought and over action.

Society and the Control of Thought

Socialization

The human being comes into this world with little other than simple reflexes, biological urges, and potential. Society makes the individual *human* in the sense of developing typically human traits: language, self, and mind. Society also socializes the human with rules and ideas that it imposes as standards for self-control. In the creation of these human traits and in the socialization of rules and ideas that become internalized in the individual, society exercises its first control over the human being: it influences *thinking*.

I am not here discussing personality traits, which, of course, are what the psychologist shows interest in. I am not posing the question of heredity versus environment in the development of personality.

Instead, I am asking questions about thinking: From where do rules come that individuals follow? From where do the ideas and values that individuals believe in come? From where does the language that they understand come? How do they get a sense of self, and how do they come to develop identity, self-appraisal, and self-control? How do they develop the ability to think with words—to talk to themselves—about the world?

The answer to all these questions is found almost entirely in the social world within which individuals are born and within which they develop. The power of society is found in the fact that in many important ways individuals are socialized into an entity that has been established before they arrive on the scene.

Language and the Control of Thought

Language use is a good example of this power. Clearly, individuals have an ability to learn language at birth and imitate sound and words very early. But the language that individuals are filled with depends on their social interaction. The words they learn—the number, type, and use as well as the importance of using words in general—depend on the society and community within which they grow up. These will become the words that they use to divide up reality, to make sense out of their world. We see and think with words. Things make sense to us through our application of the words we learn. Edward Sapir and Benjamin Lee Whorf put forth this position many years ago, and, although some think it extreme, most sociologists accept its basic outline:

> The real world is to a large extent unconsciously built
> up on the language habits of the group.... We see and
> hear and otherwise experience very largely as we do be-
> cause the language habits of our community predispose
> certain choices of interpretation. (WHORF, 1941:250)

A person in a religious community learns words pertaining to religious concerns, and then he or she is in the position to see people in that context. Some families divide the world into Christian and non-Christian; others divide it into believers and nonbelievers; others into Jewish and non-Jewish; still others into Muslim and infidel. These words become the basis for thinking about people. Some leaders remind us that the world is made of white and nonwhite people; others use more judgmental words. To enter the public schools is to learn vocabularies that open us to worlds previously unknown to us, and the university introduces us to sets of words that certain academic communities (such as sociology, physics, or psychology) continuously use. If we enter the world of a motorcycle community, the world of a gay community, or a society in Asia or Africa, we will find a new language, a new emphasis on what the community considers important in dividing reality.

It is through language that any society or community forms its dominant ideas, values, and norms. Ideas are made of the words within the language learned. The emphases in the language lead to emphases in the ideas. Thus, in a capitalist society we are likely to constantly use words such as *competition, profit, private enterprise, individual effort,* and *property.* Around these words will grow a set of ideas that is reinforced over and over. We may have ideas concerning socialism, but such ideas are reinforced less—unless, of course, they are used negatively in relation to our commitment to capitalism. The emphases in the language help create the values we hold, and they also reinforce the rules that are supposed to guide our action. How different our language would be if we lived in a monastery, a prison, the army, or on a farm, and how different would be the ideas, values, and norms emphasized.

Culture and the Control of Thought

Language is only a beginning. As we have seen, each society, community, formal organization, or group develops a *culture*—ideas, values, and norms—that is taught to people as they enter and that is reinforced through direct efforts of socialization and through people's actions. Schools, religion, television, and family cooperate in teaching the individual the culture of society and the community. Almost everything we see around us reinforces those ideas, values, and norms.

What music are we supposed to like? Although some of us think that humans freely arrive at musical taste, any analysis beyond the superficial level will show us that there are musical "eras" and that the era we happen to live in has a lot to do with what we will like. We might make some choices: if we were teenagers in the 1960s, for example, we might not have liked the Doors but might have liked the Beatles and the Rolling Stones. Today we may not like U2 but may like Nirvana or the Red Hot Chili Peppers. Yet our preference between these is minor if we realize that our choices are greatly limited by the type of music to which our generation is exposed. And look what happens to what the world around us becomes as a result. The people in our rock era develop a language with new words and phrases that others do not use or understand (*grunge, pit, feedback*) and use certain words over and over in their interpretation of reality (*rock, hot, CDs*). Each generation develops a set of ideas about music and life, and it teaches the individual certain values and norms. In the end, it influences much of what the individual thinks, and thus it influences much of what the individual does.

Indeed, when we remember that music is an industry and that an industry is meant to make money, we must look at all the ways in which the music industry influences our thinking so that we will pay for the music produced.

Music is only one example. In every area of our lives the ideas, values, and norms we come to believe in are produced in society. We learn our culture, we come to believe in its truths, but we also somehow come to believe that we have freely arrived at those truths.

For a moment, we might look at illegal drugs. What we believe and value about drugs as well as the guiding principles we follow in how to act in relation to drugs depend on the community within which we live. Think of all the social influences on us: the society we happen to live in, the community, the neighborhood, our age group, our chances for material success, the decade in which we live our youth, our school, our friends, our social class, our religious community, our academic achievements, and our experiences at work. Of course, we can "say no to drugs," but that choice is not entirely a free choice. We can also say no to selling illegal drugs, but that choice is not a free one either.

Further, our thinking about drugs is related to many other ideas we hold, all of which are heavily influenced by society. For example, we may be socialized to believe that happiness is getting high or making a lot of money, that the way to handle personal problems is to escape, or that there is nothing wrong with taking pills to solve the stress that comes with modern life. It is easy to see why so many people are attracted to drugs.

The Influence of Others on How We Think of Ourselves

We must never forget that thinking involves much more than ideas about the world we encounter outside of ourselves. We also think about ourselves. We make judgments about ourselves—we like or dislike what we are and do. Over time we develop a self-concept that is relatively stable from situation to situation. We develop identities: names we call ourselves, come to believe in, and announce to others. Such identities become who in the world we think we are.

Identities and self-judgments do not come out of a vacuum, and it is naive to believe that we freely arrive at them. To abuse a child influences that child to see himself or herself in a negative light, to see an individual who is without worth. To participate in schools where one is unable to achieve is to come to perceive oneself, at the very least, as a poor student and, in many cases, as a stupid human being. To see oneself as a doctor, lawyer, or teacher, to judge oneself as ugly or beautiful, and to regard oneself as worthy or unworthy have a lot to do with our social life. Our view of self is a result of our socialization, not just in childhood but in every social situation. It is not accidental, nor does it come from freely thinking about who we are. Our thinking about ourselves is social, just as is our thinking about the world outside.

Society and the Control of Action

I am constantly reminded of Henry David Thoreau, who, on being placed in prison for civil disobedience, declared how much freer he was in prison than other people were outside: "I saw that, if there was a wall of stone between me and my townsmen, there was a still

more difficult one to climb or break through before they could get
to be as free as I was" (1849:295). Thoreau recognized that freedom
means control: control over what one thinks. He recognized the
power of society, and he knew that most people simply accepted
what government leaders told them. He thought that his own ideas,
however, were his own, that they were relatively free of this influence. He had arrived at his own position, and through civil disobedience he had acted on that position. His action was, therefore, free
action. He controlled his thought and his action.

Now, it is very difficult to determine if Thoreau was actually
free. However, his ideas were probably a result of freer thinking than
that of others, and his action was freer than that of most others, because he controlled it, determining what he must do to act on his
thinking. But it is important to realize that once in prison, although
he was still able to think freely, he came up against a society that no
longer allowed much freedom of action to him. He could think all
he wanted, and this thinking could be free, at least to some extent.
But he could no longer act as freely as those outside the cell.

For some philosophers, such as Thomas Hobbes and Immanuel
Kant, freedom has to do with *movement*. One is free if one can move
without being controlled by forces external to oneself; action is not
restrained. When movement is interfered with, one is not free. Perhaps this is what we mean when we say "free as a bird," because
flying *seems* to be action that is not controlled by outside forces.

We therefore come to the second way in which society controls us: it not only controls what we think but also restrains us, directs us, and controls much of what we *do*.

The Control of Thinking and the Control of Action

Of course, control of thinking is the first step. To the extent that society, family, media, school, community, and primary groups socialize us to think—to accept their ideas, values, and norms—it is
difficult to see the actor as free. That is because free action means
that the actor is in control; if thinking is controlled by others, there
is little real self-control. I go to school, and I try hard to succeed: I
memorize, write and rewrite papers, and discuss the material with
others and with myself. Yes, it seems that I am in control of my own

life. But look at society: its emphasis on achieving in education (in a certain kind of education); its emphasis on a certain type of testing; its way of evaluating learning. Look at home, school, neighborhood, social class, and friends: their emphasis on education as a means for achieving material success, the individuals there who are models for me, their thinking about what constitutes a good education. All of these factors and many more influence my life and how I try to achieve happiness. And this thinking influences my *actions* in school.

In *One-Dimensional Man* (1964) Herbert Marcuse paints a picture of our modern industrial society as a place where the media dominate thinking and action and where protest becomes unthinkable. Materialism and affluence, the dominant message of the media, tell us all that is important in our lives and direct our attention to pursuing more and more wealth. Questions concerning quality of life, liberty, equality, and general human welfare are left behind. It is not that a group of conspirators decides that this must be the message. It is much more subtle: the total message, the whole atmosphere, the implied, taken-for-granted values send one dominant message, and everything else is muted.

Let us return to the discussion of our thinking about our self. What consequences this self-image has for what we do! If I think of myself as a man, janitor, or academic, then I act like one. Whether I doubt my worth or believe in myself will make a great difference in what I do. James Baldwin (1963:18), the African-American author, writes in a famous letter to his nephew: "Remember, James, ... you can only be destroyed by accepting what the white world calls a nigger." Baldwin understood that defeat comes to those who think of themselves as defeated. All of our thinking about ourselves—and thus what we do in the world—is influenced by what others say and do to us. It is difficult to hold someone responsible for his or her acts when the acts arise directly from a negative self-image fostered by interaction with others.

Institutions and Action

Actions are guided by more than thinking. We also learn how to *act* in our homes, neighborhoods, and society. We learn to follow grooves, called institutions: We marry because society has created that groove for us. We go out on dates, go steady, get engaged, have children, and get divorced because these are the various kinship institutions—or grooves—that society has developed over many years for people like us to follow. We vote from among candidates, we attend party caucuses, we go to $1,000-a-plate dinners (or, more usually, $25 bean feeds or barbecues), and we send in campaign contributions because these are the various political institutions that society has developed. We pray, take communion, get baptized, attend Christmas Mass, or attend synagogue on Saturday because these are the institutions of our community. There are economic, judicial, educational, health, and recreational institutions as well. Do we freely choose to watch television, or do we do it because this is the dominant American recreational institution today?

Socialization and Action

Socialization is the process by which we come to think and act appropriately for our society. It is accomplished through many agents: parents, siblings, friends, teachers, peers, books, movies, neighbors, clubs, gangs, the police, and employers, to name some of the most important. Realize how significant such agents are. As we act, they *reinforce* what they like, and they disapprove of acts that they do not like. We move through an obstacle course of smiles and frowns, approval and scorn, praise and anger, fines and payments, "A"s and "F"s, promotion and demotion, getting rich and going broke, gaining elective office and going to prison. Through it all we learn the directions that others approve of. We may turn our backs on them occasionally, but for most of us most of the time, the rewards and punishments matter.

Socialization depends on more than reinforcement. It also arises from *opportunities* that are made available to us. Through the acts of others we are exposed to some things in the world and excluded from others. Parents may not expose children to reading. Friends may expose them to alcohol or illegal drugs. A community may not have a respected dance school that encourages ballet. A neighborhood may have violent gangs that tempt youngsters to break the law. Socialization is not only reinforcement, but also the very subtle influence of the opportunities offered by the socializers. Our choices are restricted by these opportunities. So many of our opportunities depend on social class, ethnic group membership, and gender. If the opportunity is offered to us, it becomes a possibility for choice in our lives. If it is not offered, it is far more difficult to choose. If the opportunity is not there, if it does not exist among the people around us, how do we suddenly decide to go in that direction? And if somehow we do decide, what happens to us when we try in a world clearly weighted against our success?

Socialization also depends on *role modeling*. Clearly, human beings act in ways that are consistent with the behavior of individuals whom they respect. We dress according to what we see rock stars or favorite actors wearing. We go to school, in part, because those we consider successful go to school. We break the law because others around us who seem to be wealthy break the law. We value grades or we drop out of school, in part, because of the actions of our models. Drug use is not only a matter of reinforcement and opportunity, but also a result of what people do whom we regard as important.

The poorest children in the United States today are growing up with few role models who have steady jobs and a stable family and community life. Men are unemployed, and they feel defeated. How can children be taught that school is important in a community where few people succeed in the educational system? How can children be socialized to succeed when people around them have failed? It is easy for those of us who are not involved to blame the adults; however, they, too, are victims of a society without role models, and they, too, are victims of poverty and racism that they find almost impossible to escape.

Social Positions and Action

We are also influenced by the various positions we hold in society. We are all assigned a class position at birth, and we learn what it means to be in that class and how to act in that class. The actions we see and take on as our own are those appropriate to our class position. Interest and activity in politics are influenced by class, as is the probability of divorce or illegitimate birth. Gender-role expectations differ according to class, and so does choice of religion. Educational achievement, health care, child-rearing practices, and likelihood of criminal behavior depend on class, at least in part. Sexual behavior, dating, family life, eating, drinking, dress habits, and language—all are influenced by class.

Many of us recognize that poverty is a trap that is difficult to move out of. It focuses people's attention on bare survival, on getting enough to eat and a place to live. The focus is taken away from working for long-range dreams, on getting a high school diploma, on training for a decent job, on saving for a rainy day. Poverty means dependence on others for one's own survival. One does not generally control one's own existence if one is poor. Instead, shelter, protection, food, clothing, and medical care are in the hands of others. Clearly, the position of poverty is an all-powerful influence. If we examine those at the other extreme in society, the wealthy, we can see that their lives, too, are laid out for them: values, education, aspirations, marriage, occupation, and so on. Wealthy people usually have more control over their actions than do poor people (because the wealthy have more resources to carry out what they wish to do), but real freedom over thought and action is still very limited.

And look what gender does to us. To be a woman or a man in society is to learn a host of appropriate behaviors. Do we hold our books up against our chest or down at our sides? Do we play an assertive or a passive role in sexual encounters? Do we work to make it in the occupational world or in the world of the family? Do we major in business or in teaching? Look what it means to be a "man": virile, strong, brave, sexually aggressive, stoic, logical, practical, mechanical, dominating, independent, free, aggressive, decisive, competitive, adventurous. And we expect a "woman" to be dainty, sensual, graceful, domestic, maternal, virginal, sexually passive, flirtatious, expressive, compassionate, frivolous, intuitive, perceptive, sensitive, petty, fickle, dependent, modest, sweet, patient, affectionate, and noncompetitive (Chafetz, 1982:38–39). Of course, these may seem to be stereotypes today, and many of us do not want to believe that people hold such expectations. Yet stereotyping still influences what we become, what both men and women actually do and think. We do not have to act according to these patterns, but there are consequences if we do and consequences if we do not. In every decision we make, society tells us what is and is not appropriate for our gender.

One very interesting way to see the power of gender is to imagine another world:

> Now imagine a very different world totally void of social differences for the different genders. Families have identical expectations of sons and daughters. Children of both genders receive dolls, baseball gloves, training in how to play football, and ballet lessons. The school football team and the cheerleading squad are both coed. Books have a generous assortment of male characters behaving in ways termed feminine in twentieth-century American culture and female characters who behave in "male" ways. In all media, men and women act and dress in the same fashion. The language has been "cleansed" of all gender cues. Not only have *he, she, his,* and *hers* disappeared, but we find as many girls as boys named Joshua and Melissa. At the hospital half the nurses are male and half the doctors are female. Similarly half of all secretaries are male and half of the executives are female. How much traditional "feminine" and "masculine" behavior would survive in such an environment? (WALLACE AND WALLACE, 1985:356)

To imagine such a world and its consequences for our actions is to recognize the power of the gender role in our own society and, thus, the power of society to shape what we all become.

Roles and the Control of Action

As we saw in Chapter 5, whatever position we hold within any group or formal organization has a script attached to it called a *role*. The role dictates what the individual is supposed to do in that position, the part he or she must play. There are thousands of positions in society:

pedestrian, cook, police officer, friend, historian, quarterback, corporate vice president, television anchor, retail clerk, prostitute, server on a volleyball team, and solo soprano in a church choir. Roles tell the actor in these positions what is expected, how he or she is supposed to act. Indeed, they also tell us how we are supposed to think and who in the world we are. It is impossible to escape such roles.

To become old is to enter into a position in society that shapes the actor. Older people are supposed to withdraw from participation as workers and spouses. Retirement from work is accomplished through a series of steps: lower levels of performance are expected, other workers expect retirement, promotion ceases, and duties are reduced. It seems natural and right to retire. The elderly must also face the loss of a spouse and must be prepared to live a single life. To be in such a position brings one a set of expectations different from all other positions in society and heavily influences what we all do. It is relatively easy to see how all positions bring a different set of expectations: infant, child, preadolescent, adolescent, and so on. We expect people to act in certain ways when they are in these positions. What we all do in life depends so much on *where we are* in society.

Social Controls and Action

Finally, in considering the possibility for freedom, we must recall Thoreau's situation. Society punishes people such as him. It puts them in jail so that they are unable to act as they choose. So it is with all of our actions: they are rewarded, or they are punished. Whatever we do is subject to the social controls exercised by others on what we do.

We have prisons and fines to punish those who act outside the rules of society; we have promotions and honors for those who are good citizens. Parents scream or snicker or spank or make children feel guilty; they also talk kindly, praise them, kiss them, and make them feel good. Businesses fire, demote, threaten, and censure their employees; they also promote, praise, and give bonuses. Friends, families, groups, and communities all have sanctions—social controls—that encourage conformity.

Erving Goffman (1959) reminds us that in all interaction there are rules that we are expected to follow; and if we do not, others will sanction us. In every social situation we enter, we act a part on a stage, attempting to control how we present ourselves to others. Most of us realize that action is a performance, but if it appears too phony—that is, if it appears that the self presented is not real even to the actor presenting it—others judge the performance and the actor harshly, rejecting both. Each actor knows this and is constrained to present a believable performance. On the other hand, the others in the interaction—those who are judging the performance—are also constrained to accept most performances, to accept the self that the actor puts forth, not embarrassing him or her or revealing things that might uncover the "real" person. For we all know that if someone rejects a performance the whole basis of continuous interaction is threatened. What awaits those who try to undermine the interaction rules are negative reactions, embarrassment, rejection, and even lasting stigma. From simple interaction situations such as this to society's formal laws, human beings are governed by rules and sanctions, directing whatever they do.

Social Forces and the Individual: A Summary

The sociological perspective presents a strong case that human beings live an existence subject to all sorts of controls by other people:

1. We are located in a society and in history. Our location matters. We are all influenced by the particular social forces and social problems that exist at that point. We are influenced by other people around us, our social class, social institutions, and social groups. We do not all give in to these forces, social problems, and influences, but they still matter. At the very least we must work hard to overcome them. Living in a certain society and period of history increases the probability that we will act in certain directions rather than others.

2. Freedom is limited to the extent that society controls what we think and what we do.

3. We are all socialized by society, and this limits our freedom considerably. We learn society's language, and we learn its ideas, values, and norms—its culture. We learn how to think about the world that we encounter, and we learn how to view ourselves. Our self-concepts and our identities are created through interaction with others.

4. Our actions, too, are influenced by society. Because free action must arise from free thinking, our freedom is limited by the extent to which society creates our thinking. Socialization therefore limits our freedom considerably. Institutions, too, shape our actions, showing which way we must live our lives. Our positions, such as social class, form our actions, as do roles. Finally, people exercise social controls, directing us in the desired directions.

Is There Any Freedom?

Remember that I began this chapter by asserting that it is impossible to determine whether human beings are free. It should be obvious that much of what we do is a result of society. We are shaped and controlled by many forces. But is there anything left? Are we in any way free? Can we point to people and, in truth, declare: "They are responsible for what they do"?

Most sociologists do not take an extreme position on freedom. As important as society is, there is something about human beings that makes us more than robots in society. All of us overcome the efforts of the socializers to some extent, and some of us do so to a remarkable degree. We are, at least to some extent, "active" in our environment, shaping our world, acting in relation to that world, and making decisions about what to do in that world.

If we are to understand the possibility for freedom in society, we must broaden our appreciation of the importance of society for what we all become. Socialization not only brings social control but also gives us the ability to *think* about our actions and the actions of others. Humans are, Max Weber emphasizes, beings who give meaning to their actions. In other words, they *understand* their own acts and the acts of others. They interpret the world, not simply respond to it. Language (derived from socialization) is the tool we employ to think about our situation, to consider options, to appraise the morality and the effectiveness of our own actions, to consider the future and apply the past, and to understand other people's thinking and feelings. This ability to think with words helps us to break out of the simple social conditioning that I have described thus far in this chapter.

No sociologist denies the power of socialization, structures, culture, and institutions. But once we add the ability to think, there is some room for free choice. If we define freedom as the ability of the actor to control his or her own life, this ability to think is a prerequisite. Most of those who claim that humans are free make reflection a central aspect of that freedom. Steven Lukes (1973:52) describes the autonomous individual as one who "subjects the pressures and norms with which he is confronted to conscious and critical evaluation, and forms intentions and reaches practical decisions as the result of independent and rational reflection." I make decisions, I choose, I consider alternatives. This rational relationship to the environment and to our own action is central to what it means to be free. Instead of simply being shaped by others, we tell ourselves what must be done. And even if we do conform to society, conformity, too, can arise out of such reflection and choice.

Goffman (1961) describes active humans playing roles, choosing, evaluating themselves and others, distancing themselves from some roles, shaping roles, figuring out situations, and performing in them. Gary Fine (1984) reminds us that roles are negotiated, that acts are not fixed but arise out of the give-and-take of social interaction, worked out as people act back and forth. Society is *active drama*. Humans, because of their socialization, take at least some of the control from the social world, and they determine to some extent how they are to think and act. This is what George Herbert Mead (1934) means by *self*: to possess a self means that the individual is *in control*. No matter how important the social environment, such individuals are able to temper that influence somewhat and to tell themselves what to do.

No sociologist regards freedom as involving individuals simply acting out of impulse, uncontrolled, untouched by society, doing whatever they feel at the moment. Freedom is not the absence of social organization, social control, and socialization. It is, instead, self-direction within the larger context of society. To end organized life is to end human life. Durkheim persuasively argues that freedom for human beings can exist only in the context of a social, moral order. It is self-control within a broad framework of social control. Without social organization there would be chaos, and with chaos there are no rules. As romantic as this might sound, without rules it is hard to conceive of anything but social conflict, power based on force, and a disregard for the freedom of anyone but oneself.

Perhaps one way to see the possibility for freedom in society is simply to recognize that all social organization sets broad limits within which we all act and within which we have some freedom to make choices. Organization cannot exist if it is rigid, unyielding, and in total control of the actor. Thus, although we have a society that tells us that to get ahead we must go to college and try to get a practical education, each of us has some choice in which specific major to work in. Although we are told that it is important to marry and are influenced to marry certain kinds of people, we exercise some freedom to choose a mate. Although everything seems to tell us to get a college degree, find a major that will guarantee a secure financial future, and take certain classes, we still have some choice among two or three colleges, among two or three majors, and among two or three classes to take.

This does not mean that freedom always includes obedience to society as it is. On the contrary: freedom certainly includes a critique of society and thoughtful actions to alter or even overthrow it. However, criticizing, changing, and overthrowing society does not bring the absence of organization but an alternate organization, either more encouraging of individual freedom or more stifling.

This discussion actually leads us to one final consideration. For some of us a great deal of free choice is possible; for others choice is very limited. Financial security generally allows for freer choices than does poverty. Control by one group affords the individual less freedom than participation in several groups. Power also matters: if power is the ability to achieve one's goals in relation to others, those without power will be unable to achieve their goals. They are powerless, they are dependent on others, and their lives are not their own. If one is poor, then in relation to others one is normally not very free. If one is wealthy, one is normally freer. If one lives in a society where there is a totalitarian state, one is not very free. If one lives in a society where ideas are constantly being espoused, exchanged, and argued, one is normally freer. If one lives in a society where conformity is highly valued, it is difficult to be free; if diversity is valued, freedom is easier. Always, however, there are limits, and the more one understands about the human being, the greater these limits seem to become.

Summary and Conclusion

Sociology is not a simple perspective; its answers are always complex. Social controls over the human being are forever present and place significant limits on freedom. It is naive to deny their existence; indeed, it is essential to understand their importance if we are to understand who we are and why we act and think as we do. Yet all of us have some freedom; this is impossible to prove, as I have said, but it seems likely if we consider the nature of the human being. The sociological approach to the possibility for freedom goes back to a simple but very profound theme: *language, thinking, and selfhood are basic to freedom, and all of these are socially derived qualities*. They arise in interaction. They are a central aspect of our socialization.

References

The following works deal with the question of freedom in society by examining the possibility for freedom as long as human beings live in society. Some show more directly how much some of us are determined by our social environment.

Adams, Bert N. 1979 "Mate Selection in the United States: A Theoretical Summarization." In *Contemporary Theories About the Family*, Vol. 1. Ed. Wesley R. Burr et al. New York: Free Press.

Auletta, Ken 1982 *The Underclass*. New York: McGraw-Hill.

Baldwin, James 1963 *The Fire Next Time*. New York: Dial Press.

Bellah, Robert N., Richard Madsen, William M. Sullivan, Ann Swidler, and Steven M. Tipton 1985 *Habits of the Heart: Individualism and Commitment in American Life*. New York: Harper and Row.

Berger, Peter 1963 *Invitation to Sociology*. New York: Doubleday.

Berger, Peter L., and Thomas Luckmann 1966 *The Social Construction of Reality*. New York: Doubleday.

Blumer, Herbert 1969 *Symbolic Interactionism: Perspective and Method*. Englewood Cliffs, NJ: Prentice-Hall.

Chafetz, Janet Saltzman 1982 *Masculine/Feminine or Human: An Overview of the Sociology of Gender Roles*. Itasca, IL: Peacock.

Charon, Joel M. 1995 *Symbolic Interactionism: An Introduction, an Interpretation, an Integration*. 5th ed. Englewood Cliffs, NJ: Prentice-Hall.

Clausen, John A. 1986 *The Life Course: A Sociological Perspective*. Englewood Cliffs, NJ: Prentice-Hall.

Durkheim, Emile 1895 *The Rules of the Sociological Method*. 1964 ed. Trans. Sarah A. Solovay and John H. Mueller. New York: Free Press.

Durkheim, Emile 1915 *The Elementary Forms of Religious Life*. 1954 ed. Trans. Joseph Swain. New York: Free Press.

Eliade, Mircea 1954 *Cosmos and History*. New York: Harper and Row.

Ewen, Stuart 1976 *Captains of Consciousness*. New York: McGraw-Hill.

Feagin, Joe R. 1975 *Subordinating the Poor: Welfare and American Beliefs*. Englewood Cliffs, NJ: Prentice-Hall.

Fine, Gary Alan 1984 "Negotiated Orders and Organizational Cultures." *Annual Review of Sociology, 10*:239–262.

Freud, Sigmund 1930 *Civilization and Its Discontents*. 1953 ed. London: Hogarth Press.

Fromm, Erich 1962 *Beyond the Chains of Illusion*. New York: Simon and Schuster.

Goffman, Erving 1959 *The Presentation of Self in Everyday Life*. New York: Doubleday (Anchor).

Goffman, Erving 1961 *Asylums: Essays on the Social Situation of Mental Patients and Other Inmates*. New York: Doubleday (Anchor).

Hertzler, Joyce O. 1965 *A Sociology of Language*. New York: Random House.

Jacobs, Nancy R., Mark A. Segal, and Carol D. Foster 1988 *Into the Third Century: A Social Profile of America*. Wylie, TX: Information Aids.

Lukes, Steven 1973 *Individualism*. New York: Harper and Row.

Marcuse, Herbert 1964 *One-Dimensional Man*. Boston: Beacon Press.

Marx, Karl, and Friedrich Engels 1848 *The Communist Manifesto*. 1955 ed. New York: Appleton-Century-Crofts.

McCall, George J., and J. L. Simmons 1978 *Identities and Interactions*. New York: Free Press.

Mead, George Herbert 1925 "The Genesis of the Self and Social Control." *International Journal of Ethics*, 35:251–277.

Mead, George Herbert 1934 *Mind, Self and Society*. Chicago: University of Chicago Press.

Milgram, Stanley 1963 "Behavioral Study of Obedience." *Journal of Abnormal and Social Psychology*, 67:371–378.

Mills, C. Wright 1956 *The Power Elite*. New York: Oxford University Press.

Mills, C. Wright 1959 *The Sociological Imagination*. New York: Oxford University Press.

Perrow, Charles 1986 *Complex Organizations*. 3rd ed. New York: Random House.

Shibutani, Tamotsu 1961 *Society and Personality: An Interactionist Approach to Social Psychology*. Englewood Cliffs, NJ: Prentice-Hall.

Thoreau, Henry David 1849 "On the Duty of Civil Disobedience." In *Walden: On the Duty of Civil Disobedience*. 1948 ed. New York: Holt, Rinehart and Winston.

Wallace, Richard Cheever, and Wendy Drew Wallace 1985 *Sociology*. Boston: Allyn and Bacon.

Warriner, Charles K. 1970 *The Emergence of Society*. Homewood, IL: Dorsey Press.

White, Leslie A. 1940 *The Science of Culture*. New York: Farrar, Straus and Giroux.

Whorf, Benjamin Lee 1941 "Languages and Logic." *Technology Review*, 43:250–252, 266, 268, 272.

Wilson, William 1987 *The Truly Disadvantaged: The Inner City, the Underclass, and Public Policy*. Chicago: University of Chicago Press.

Wrong, Dennis H. 1961 "The Oversocialized Conception of Man in Modern Sociology." *American Sociological Review*, 26:183–193.

Zimbardo, Philip 1972 "Pathology of Imprisonment." *Society*, 9:4–8.

Why Can't Everyone Be Just Like Us?

The Dilemma of Ethnocentrism

*T*he ancient Greeks lived in many small city-states. Each city-state was independent. Each had its own government, army, and economy. Some, such as Athens, were democracies, and some, such as Sparta, were autocracies. Together, however, their citizens shared a heritage: they were all Greeks. Beyond the mountains and sea lived *other* peoples. Such people were strangers, barbarians. Their ways were different and less desirable. They were, in a word, "uncivilized." Like the Greeks, the Romans also saw the world divided into two: the civilization of Rome and the barbarian peoples. The medieval world divided people into heathen and Christian, and the European peoples who came to the Americas encountered many different cultures but called them all "Indian" and commonly described their ways as savage.

When I attended North High School, I honestly believed that somehow our school, our student body, our teachers, our teams were better than others. My loyalty to North included a lingering belief that we were truly blessed over those who attended other schools. At all athletic events I was sure that in controversial calls we were right and the other team somehow had the referees on its side.

Most of us do, in fact, believe that the United States is the greatest nation in the world, and it is difficult for us to believe that there are other ways of living that are equally good or even better. When we looked at the former Soviet Union, we blamed its problems and shortcomings on an authoritarian regime and on government involvement in the economy, and we claimed that if only the Kremlin's ways could become similar to our ways, the people could have enjoyed what we enjoy. Indeed, when we look at other cultures, we tend to distinguish them according to how close they come to our own: we see some as primitive, some as developing, and some as developed and civilized.

I am talking here of several very important issues, all intimately related. It is important to examine these one at a time.

The Meaning of Values

"Oompa, oompa, oompa-pa, my pa's better than your pa-pa." My religion is better than yours. My school is. My major is. My parents are. My morals are. My life plans are. My goals in life are. My car is. My friends are.

Comparisons have something to do with *values*. Whenever we use terms such as *better, best, good, bad, superior, inferior, should*, and *should not*, we are entering the complex world of values. The tipoff to me that people are discussing values is whenever they use or imply the word *should*. In that case someone is always making a value judgment. The statement has to do with what should exist in the world rather than what actually exists. The title of this chapter is "Why Can't Everyone Be Just Like Us?" Although *should* is not in the question, it is certainly implied: "Others should be like us, so why aren't they?"

I vividly remember a conversation with two professors on a four-hour drive. They were singing the praises of higher education. "Everyone should get a college education," they said. "Knowledge is better than ignorance." I turned to them and boldly declared: "You're making a value judgment. Although I generally agree with you, there's no way any of us can prove that we're right. Only statements of fact can be proved." They disagreed, and we argued back and forth. I asked: "Why is knowledge better than ignorance?" Their answer: "Because it helps us succeed in the occupational world." "Well," I replied, "who says that we *should* succeed in the occupational world?" Such questions are often important and sometimes trivial, but they always involve assumptions of what we think life *should be like*, and thus they become questions of values.

Values are our commitments, and they reflect our image of what is good and what is not good in this world. Values are the standards against which people judge their own acts and the acts of others. They tell us "what goals people ought to seek, what is required or forbidden, what is honorable and shameful, and what is beautiful and ugly" (Shibutani, 1986:68). If I really believe that having a family is important to a meaningful life, that is a value to me. I live my life for my family, I vote on issues affecting my family, and I spend time and money on my family. Perhaps I even broaden this commitment to acting in favor of family life throughout the United States and even the world. As often happens, I become so committed to my family that I find it difficult to understand how others who do not have a family life similar to mine can possibly find happiness. I may also claim that this lack makes them immoral or selfish. I find threats to my family and family life in general to be important threats to my existence, and I support efforts to rid society of these threats.

For some of us, freedom is an important value ("I should be free, all Americans should be free, and all people should be free"). Likewise, law and order might be a value, or religion, equality, artistic expression, education, a healthy body, physical beauty, tradition, individualism, friendship, helping others, living a moral life, making a lot of money, being a good citizen, and so on. These are all examples of what we regard as worthwhile. If I believe in them, they are my values; if you believe in them, they are yours. But there is no way either one of us can prove that ours are better than the other person's, for whenever we try to do this, we inevitably come up against more and more value judgments, none of which can be proved.

Our values can be contradictory. Americans can believe in both a segregated society and equal opportunity for all, or they can find themselves simultaneously worshiping individualism and group loyalty. I find myself attracted to tradition and progress at the same time, and sometimes I am torn between spending my time writing a book and listening to the concerns of my wife and children.

Most of our decisions in life involve choices we make between several values that we hold. We might value both freedom of expression and the rights of women. On the issue of pornography, we might have to choose between these values: "Yes, I value freedom of expression, but I don't think people have the right to produce pornography that denigrates women." Or on matters of civil rights: "Of course, I favor equality for all races. But I also believe that people should go to schools in their own neighborhoods." It is not always easy to turn our backs on one of our values so that we can work for another, but on occasion we must do just that. And, of course, this causes conflict in most of us whenever we recognize the contradiction.

Judging other people is how values enter into our social life. We like others on the basis of value judgments we make: "He is a true individual." "She is a really ambitious person." "I respect the fact that he speaks up." "She is really pretty." "What a nice guy." We also dislike others based on value judgments we make: "He's dishonest." "She's stuck up." "They're immoral." "They're lazy." "They're barbaric." In every case we create a measuring stick (a value) and use it to judge others. And, of course, judging is more than just liking or not liking others. It is also deciding who should be punished, who should be promoted, whose death is called for, or who should live a happy life. It is deciding who must be changed, and whom we must make war on.

When we ask, "Why can't others be just like us?" we are asking a question based on a yardstick we have somehow developed. It is a question that is at heart a statement of values, a statement that our ways are better than others' ways and that to make a better world, others should become like us.

All of us probably do this type of judging on occasion. Some of us do it often. But value judgments are statements of preference, not fact. There is no way to prove that "my pa's better than your papa" unless we specify what we mean by better, and as soon as we do that, we are making a value judgment, which really is an *assumption* about what is preferable in the world.

The Meaning of Ethnocentrism

All organization is established through interaction. Social patterns—structure, culture, and institutions—are established over time. On top of these patterns there usually emerges an attraction to the organization, a feeling of loyalty, a sense of togetherness, and a commitment to one another. Such a feeling is encouraged simply by the interaction itself and by the intentional efforts of individual actors. Of course, this interaction can end at any point if common goals are achieved, if boredom sets in, or if destructive conflict pulls people apart; so continuation of interaction is always tenuous.

For a moment, however, let us imagine what happens as people interact on a continuous basis. Each acts with the others in mind; people act back and forth; each considers the acts of the others. The more this interaction occurs, the less each has an opportunity to encounter outsiders. If interaction is lessened, there is more time to go elsewhere. Going steady is a case in point: it usually means cutting off regular interaction with other people of the opposite sex. Going steady sometimes means cutting off regular interaction with friends of the same sex, too. One relationship takes over, and time is short for others. It is difficult to have close friendships with many people simply because a close friendship takes time and commitment; the more it takes, the less time we have to develop others.

Continuous interaction tends to develop a likeness among the actors. Actors communicate, they share and discuss experiences, and they adopt rules, ideas, and values. They devise ways of dealing with the world they encounter. They develop a language that has a unique meaning to them. In short, they develop *culture.*

At the same time, continuous interaction isolates actors from outsiders. Differences with those outside the interaction are created and accentuated, communication does not occur regularly, and sharing ideas proves more and more difficult. In time, actors develop a set of meanings, understandings, and values different from those of outsiders.

What happens, of course, is that outsiders not only appear different to us, but also come to be seen as strange, maybe deviant, maybe ill or evil. We make value judgments on the basis of what we are familiar with. We judge others on the basis of the world within which we interact. It is common for us to develop what sociologists call ethnocentrism.

Ethnocentrism means that people think that their culture ("ethno") is central ("centrism") to the universe. It is a tendency to use what we have shared—values, ideas, and rules—in interaction as a starting point for thinking about and judging other people. It is possible to see other people as different without making value judgments, but it is difficult. We tend to think in terms of what we have learned in interaction, and it is hard for most of us to stand back and declare: "They are different. So what? That doesn't mean they are any worse than we are. That doesn't mean their ways were developed as a critique of our ways."

Think of ethnocentrism existing throughout our social existence, from the smallest group we join to the society we live in. Actually, we might even imagine the earth as a whole and think what would happen if we encountered a world with different beings. In every case we see interaction, sharing, isolation, differences from outsiders, and the tendency to develop feelings of ethnocentrism. Not all individuals fall into this trap, but virtually every social organization does. Why? Why is there such a strong tendency for people in a social organization to make value judgments about people outside that organization and declare: "Why can't everyone be just like us?"

The Reasons for Ethnocentrism

Social Interaction Encourages Ethnocentrism

Ethnocentrism develops, first of all, simply because of the nature of *interaction*. We interact and share; we become organized, form a structure and institutions, share a culture, and, as a result, tend to become isolated from others with whom we do not interact. Groups develop differences from one another; formal organizations do; communities do; societies do. Without interaction with outsiders, differences become difficult to understand and very hard not to judge. What is real to us becomes comfortable, and what is comfortable becomes right. What we do not understand becomes less than right to us. Ethnocentrism is encouraged.

Loyalty to an Organization Encourages Ethnocentrism

Ethnocentrism also develops, however, because of the nature of *social organization*. As we interact and become part of a society or a group, we generally come to feel something good about belonging to that group. We are not only American, but also we come to feel *good* about being American. We support our troops in the world; we tend to give our leaders any benefit of the doubt when there is conflict with other nations. Our identity becomes tied to what we feel good about. Life takes on meaning in that organization. We feel good that we belong to something. I am a marine, a Xerox employee, an Elk, a New Yorker, a student at Harvard, a member of the National Association for the Advancement of Colored People. Belonging brings direction, comfort, and security. Belonging brings a social anchor to our lives, giving meaning to what we do and more certainty to what we believe. Becoming part of a social organization (from a small group of friends to a large society) encourages a *sense of loyalty*, and that loyalty encourages ethnocentrism, for loyalty means a commitment to something we regard as important and right. Loyalty brings a feeling of obligation to serve and defend. That, of course, means that criticism and threats to the organization are defended against. *It is easy to see alternative ideas, values, rules, and actions as threats to what we feel loyalty to rather than simply qualities that are different from ours.* This is a basic cause of ethnocentrism.

The defenders of an organization—be it a society, community, formal organization, or group—encourage ethnocentrism by what they teach. To the extent that loyalty to the organization is stressed, ethnocentrism flourishes. Rituals are developed and function to bring people together and reaffirm what is right—that is, the organization. People are encouraged through song, secret ceremonies, flags, dress, and statements of faith to feel that their identities are tied to the organization: "I am an Iranian." "I am a Lutheran." "I am an Owl." "I am an IBM employee." "This is who I *am*. This gives my life meaning." It is a lot easier for defenders of society to ensure cooperation through such loyalty than through any other means. Teaching such loyalty, in fact, encourages people to see their organization as right; it generally contributes to ethnocentrism.

Of all the sociologists who have investigated this phenomenon of loyalty, no one is as important as Emile Durkheim. One of the ways in which loyalty is learned and felt is through representations of the community, which Durkheim calls "sacred objects." Religious symbols are sacred objects (communion wine, a cross, a Bible, a costume, a church, a holy person). Certain political symbols are also sacred objects (a flag, a tomb, an anthem, a historical document, certain political offices). And some people regard various other objects as sacred (Elvis Presley's jeans, Babe Ruth's bat, Liberace's piano, and Martin Luther King's "I Have a Dream" speech). Sacred objects take on a meaning significantly greater than their physical nature. A cross, for example, is not simply an artistic design or two pieces of wood tied together. On the one hand, such objects help bring the community together, because they are believed to represent the community and what it stands for. Whenever they are produced, people feel united, they feel good about being part of the community. On the other hand, such objects take on a *sacred* quality; they help make the group (actually a social organization) into something that is right and above the ordinary (a sacred community). Loyalty is combined with the feeling that what is believed in the group is more than social. It is true, right, universal, and above the ordinary or mundane: it is sacred. Through the creation of sacred objects people are taught that their world is right, and this lesson becomes embedded in their emotional life.

Socialization Encourages Ethnocentrism

Ethnocentrism thus arises as people are encouraged to feel loyalty to an organization and as they learn through sacred objects to feel that their community and its culture are special. Ethnocentrism is also developed as people come to believe that their ideas, norms, and values are sacred.

   Every society has morals, laws, values, and beliefs that it socializes its people to believe. But why should we believe them? Because someone else teaches them? Because others around us believe them? Because we are told we will be rewarded if we do and punished if we don't? Not good enough! Order in society depends on rules and beliefs that people somehow regard as special, right, and true, not rules and beliefs that are simply agreed on socially. Teachers, political and religious leaders, and families cooperate to demonstrate to the individual that society's ways are more than social, that they are universally true and good. They are part of a sacred order rather than simply a social order. Ethnocentrism must, in fact, result, for if our society possesses what is true and good, it becomes difficult to believe that societies that are different are also true and good. The tendency is for people to see truth and goodness in others to be that which comes closest to what they consider to be true and good.

We can see this tendency in every group, not just societies. In an organization's attempts to teach its members what is true and right, there is a tendency to teach the universality of these values, leading to the belief that different values are less worthy.

The Creation of Deviance Encourages Ethnocentrism

Where there are rules, there will always be violations of those rules. Lines are drawn, and people are punished. Punishment shows all members of society that individuality can go only so far. It shows them the consequences of violating rules.

The condemnation of some people both inside and outside an organization also encourages ethnocentrism. Sociologists define *deviants* as those individuals who are perceived to be violators of the rules and truths of society, and are therefore perceived threats to social order. They are "outsiders" in that they are outside of what good people know is true and right. They are barbarian, uncivilized, savage, criminal, insane. Each society creates its own outsiders by drawing lines: "Over this line there is something *wrong* with you." The lines shift, but they always exist. To accept widespread individuality is to admit that there is really nothing special about the truths, values, and rules we all believe in. It is to encourage individuals to question, to see the social basis for much of what they believe rather than a more absolute basis. All groups must therefore draw lines: "Outside this line your beliefs or actions are unacceptable and will be condemned." The creation of such lines, the punishment dispensed to those who cross those lines, and the lasting stigma attached to those who are punished are ways in which society reaffirms its rules, reinforces the sacredness of those rules, and, simultaneously, creates ethnocentrism. In a way, punishment creates throughout the community a greater certainty that "we are right."

Dominance and Oppression Create Ethnocentrism

The slave trade that prospered from the seventeenth to the early nineteenth centuries was the product of people who realized that there was a fortune to be made by uprooting, transporting, and oppressing large numbers of people without any concern for their own desires, plans, values, or ways of life. Like most of the rest of humanity, slave traders and slave owners probably believed in God, and they probably considered themselves good, upstanding citizens of their world. It is too easy for us to dismiss them as insane or evil. How, in fact, did they live with themselves? Did they have conscience? Did they consider themselves to be moral people?

Cases of inhumanity exist in every people's history. The United States systematically destroyed Native Americans. The Germans murdered millions of people who were defined as less human. Southeast Asia and Yugoslavia in the 1980s and 1990s are filled with more examples of one group of people systematically and intentionally killing others whom they define as different.

What is the link between such oppression and ethnocentrism? Without question, ethnocentrism sometimes encourages war, systematic murder, slavery, exploitation, and inequality. Yet it is also true that ethnocentrism is the result of such acts. Racism did not precede (and cause) slavery; it is clear from the historical record that it was the existence of slavery that influenced the development of racism. Slavery was developed for economic gain, not because one group was seen to be inferior. A racist philosophy in turn developed to try to justify and protect that institution.

Extend this argument to any instance of inhumanity. Where people oppress others, there normally needs to be a justification for their actions to convince themselves and others that what they do is all right. Some form of ethnocentrism is generally the result. It is all right to oppress, because "what they are" is less worthy than "what we are." "God decided that our people should conquer and control the world. The sacrifice of others for our benefit is both necessary and right." The "old boys network" develops a rationale for its treatment of women that tells them that their own ways are superior; the employer who exploits cheap labor comes to believe that he or she is helping "those people" who do not need the same income as "people like us"; and the conquerers who grab the land and imprison or destroy those who owned it explain that "they didn't use it the right way anyway."

Recognize, then, that ethnocentrism is an ideology, a way of thinking that people use to justify to themselves and to others the oppression of people unlike themselves. Where oppression exists, ethnocentrism is encouraged.

Social Conflict Encourages Ethnocentrism

Social conflict is an inherent part of all social life. Wherever there are differences or wherever there is scarcity, there is conflict, not necessarily violent conflict but at least a struggle over whatever is scarce. Interorganizational conflict (conflict *between* organizations) normally encourages ethnocentrism. War is the best example, but less violent competition between companies, teams, or communities also reveals this tendency. Those with whom we do battle are portrayed as less worthy and as deserving of our contempt:

> We tend to impute to our enemies the most foul
> motives, often those that we have trouble avowing our-
> selves: [we tend to believe that] the enemy is inherently
> perfidious, insolent, sordid, cruel, degenerate, lacking in
> compassion, and enjoys aggression for its own sake.
> Everything he does tends to be interpreted in the most
> unfavorable light. (SHIBUTANI, 1970:226)

At the same time, when we are involved in conflict we tend to describe our own motives and our own ways as noble.

> [We maintain that] we seldom engage in wars because
> of greed. We fight for freedom and injustice or in
> defense against unwarranted aggression. We are strong,
> courageous, truthful, compassionate, peace-loving, and
> self-sacrificing. We respect the independence of others
> and are loyal to our allies. (IBID., 226)

In conflict with others, we tend to idealize our own ways. We selectively see who they are and who we are. We exaggerate the faults of the other; we exaggerate our own virtues. Enemies are transformed from human beings into objects without rights, and it becomes increasingly difficult to see the world from their perspective. What we do against the enemy becomes more acceptable to us, because we are able to rationalize the defense of goodness against evil. Conflict leads us to increase our ethnocentrism, and increasing our ethnocentrism serves to justify and increase conflict. The two build on each other, and over time the world appears to be more and more obviously a struggle between good and evil. The 1990–1991 crisis in the Middle East, for example, increasingly became a struggle between the "forces of good," as represented by President George Bush, and the "forces of evil," as represented by President Saddam Hussein. Georg Simmel (1908) reminds us that interorganizational conflict not only encourages group loyalty but also exerts pressures for conformity to silence internal dissent. "Moderate and reasonable men are virtually immobilized, and the public gets a constant repetition of a single point of view" (Shibutani, 1970:228). In time of war, a society is most sensitive to criticism among its own population. Such criticism is perceived as disloyalty, just as the social conflict with outsiders is perceived to be a threat to everything that is right. It is much easier to tolerate dissent when the society is at peace. Social conflict brings out the rightness of our cause, our ways, our truths. It is difficult for participants to stop this cycle of social conflict, increasing ethnocentrism, and persecution of minority opinion once it has begun. Gang fights, wars, and cutthroat business practices are often the result. In short, interorganizational conflict encourages people to believe in the rightness of their cause, their ways, their truths.

Summary

Ethnocentrism is very common. We have looked at five reasons for it, most of which are embedded in the very nature of social organization:

> 1. Interaction itself encourages ethnocentrism, just because it causes us to be around certain people, to be isolated from other people, and to become acquainted with certain views and not others. Over time, those with whom we do not interact appear different, strange, and, usually, less worthy.

2. As people develop feelings of loyalty to an organization, it is difficult to escape becoming ethnocentric. Loyalty develops as we feel commitment and is encouraged by those who represent the organization. Ritual and sacred objects not only socialize people to feel loyalty but also tend to make the organization sacred in the minds of its members.

3. Much of our socialization is an attempt by others to get us to believe that the rules and ideas they teach are right and true rather than simply social, agreed on, useful, and worldly.

4. We punish violators of our rules, and in so doing we declare that other rules are unacceptable.

5. Oppression encourages ethnocentrism, which serves the oppressor by justifying actions made against others who are defined as less worthy people.

6. Ethnocentrism is encouraged when there is conflict with another social organization. Conflict prompts us to denigrate the other side and to develop more extreme degrees of ethnocentrism to justify our actions. Silencing dissent in times of conflict results from equating criticism of society with disloyalty; conformity and ethnocentrism are equated with loyalty.

Human Differences

"Why can't everyone be just like us?" The question usually involves a value judgment. ("After all, our ways are better!") The question therefore is ethnocentric. ("Our ways are right; ways that differ from ours are less right.")

However, let's ask the question without making a value judgment, without being ethnocentric. Let's become more scholarly and objective and phrase the question differently: "Why are people different from one another?" "Why are societies different?" "Why are groups and formal organizations different?"

Social Interaction and Human Differences

We should already know part of the answer. Earlier in the chapter I pointed out that interaction brings differences. As people interact with one another, they are pulled together and, at the same time, separated from others. Interaction means that people share, they become familiar, they increasingly depend on one another, they form social patterns—culture, structure, institutions—and they become used to what they have formed. Over time their world becomes part of them, and it appears to be a natural part of the universe.

It is this interaction and the patterns that result that distinguish people from one another. My ideas, values, morals, and traditions are different from yours because the groups within which we were formed are different. I have been influenced by different social organizations. Drugs and alcohol are not part of my life because my interaction has not taken me in that direction. The religious beliefs I hold are traceable to my interaction, and so are my interests and talents. My life is different from yours in part because I grew up in Minneapolis and moved to Moorhead, and you stayed in Minneapolis. My life is different from yours because I interact with sociologists and your contacts are different.

The real meaning of social organization is that it brings commonality, communication, and cooperation with those inside, and it also brings differentiation from, lack of communication with, and much less cooperation with those outside. Interaction and organization bring internal unity and external differences. So long as there is interaction and so long as that interaction does not include everyone at one time, it is impossible for all of us to be the same.

Social History and Human Differences

No two societies (or groups or formal organizations or communities) develop in the same way. All have a different history. The unique aspects of their development will produce differences between those inside and those outside society, making it impossible for "them" to be like "us." Societies may appear to be alike in that all have important charismatic leaders in their history—a Lenin, a Luther, a Muhammad, a Napoleon, or a Gandhi—but each one of these leaders will have brought a unique set of changes, unlike the leaders in other societies. Each society will have a mixture of tradition and modernization, and that mixture will always be unique. Each society will be at a different stage of industrialization, and each will combine that industrialization with a different heritage. Each may depend heavily on one major religion, and in that sense they will be similar, but each religion will be different in several basic ways. Even when two have the same religion (Catholicism, for example), each will, in fact, be different, because the religion will exist within a larger social context. Iran, Syria, Egypt, and Indonesia may all be Muslim, but the life, ideas, values, and even the religion of the people will differ considerably because of their different histories.

All social organizations have unique histories and thus create different social patterns from all others. I belonged to both a poker group and an investment group in Minneapolis. When I moved to another city, I helped form these groups anew. The groups in Minneapolis still exist; in Fargo-Moorhead they also exist. Even though I tried to form the same groups that existed in Minneapolis, they evolved much differently. Why can't the groups in Fargo-Moorhead be just like the ones in Minneapolis? Because these groups differed in their history: their experiences, problems, solutions, and social patterns.

"Why is it so difficult for African Americans in the United States to make it? Other minorities have. Other immigrant groups have. The Jewish people have. The Japanese Americans are. What is different about the African Americans?" The answer to this question is complex, but part of it is found in the different social histories. It lies in the history of the United States (in the force that brought Africans to the Americas, in the institution of slavery, in the Civil War, in post–Civil War conflict and domination, in the immigration of large numbers of whites in the late nineteenth and early twentieth centuries, and in the migration of African Americans from southern rural areas to northern urban areas during and after World War I). It lies in the nature of the historical relationship between whites and African Americans (the patterns of segregation, poverty, exclusion, and domination that prevailed for hundreds of years). It lies in our interaction patterns in a highly segregated society, which led to separate communities without open interaction and communication, encouraging different and sometimes clashing social patterns. It lies in a heritage of mistrust and hopelessness, fostered by discrimination in every area of American life. African Americans are different from every other minority in our history; so, too, are Jewish Americans, Japanese Americans, Native Americans, and Mexican Americans. There is no reason to believe that these groups are the same simply because they are or were disadvantaged.

Problems and Social Patterns

People therefore differ from one another because their interaction separates them and because their unique histories create different social patterns. Groups, formal organizations, communities, and societies also develop differently from one another because the problems they encounter are different. Organizations develop structure, culture, and institutions that *work*. China cannot be like the United States because the problems it must solve are very different from those in the United States and therefore call for different social patterns. For example, the problem of unity and social order has always plagued China. China has really been many societies, not one, and there has been a strong tradition of division. The history of China is one of separate feudal empires, fighting warlords, decentralized governments, and decentralized economies. In contrast, the United States, although it began as separate states, has a stronger tradition of unity, fostered by the Revolutionary War and the founding of a society separate from England, forced by the Civil War, and encouraged by transportation, communication, and economic systems that developed rapidly after the Civil War. China also has been conquered by Japan and attacked by the Soviet Union and has developed a mistrust of its neighbors. This fact influences its ways. The United States, on the other hand, has never lost wars to its neighbors and has not developed this same fear. Finally, the massive population problems created out of a long history of loyalty to family and tradition as well as a long history of widespread poverty have made China a different society from the United States.

Given these different problems, how can we think that the U.S. and Chinese societies can be alike? How can we imagine that what works here is going to work there? Private enterprise may be a great institution in the United States; it is difficult to transplant it to a society with different problems to deal with that has traditionally valued kinship and community over individualism.

Baseball teams that win pennants cannot be like baseball teams that are trying to win their first game. Universities that graduate those who fill elite positions cannot be like universities that try to offer some education to everyone who wants it. Communities that have serious pollution problems cannot be like communities that must solve the problem of unemployment.

We are often tempted to compare ourselves with other societies, bragging about our progress or even complaining about something we would like improved. It is common for those of us in education to yearn for the educational system of a Great Britain or a Singapore. However, the purpose of our educational system has always been different from that of these other societies. We have tried to build a high school system that works to equalize opportunity as much as possible and a university system that tries to appeal to the needs of the entire population. Our resulting institutions have therefore been different. For good or for bad, ours is an open system, in which we give the individual many opportunities for success, and until we radically change the purpose of our schools, it is impossible to build a school system similar to those in other societies. If we accepted only the most academically talented in our high schools and then closed the universities to all but the most talented, we could develop a system of education similar to those in other societies, but the whole purpose of education would change and with it the whole nature of our society. Institutions do not develop in a vacuum. Our ways have developed around values and problems that we have designated as important.

Simply put, organizations differ from one another for three reasons: (1) interaction isolates and differentiates them, (2) their histories are unique, and (3) the problems with which their social patterns must deal are different, and this influences what patterns develop.

A fourth reason should be pointed out. Remember that earlier in this chapter we explored ethnocentrism, the tendency for us to regard our ways as right and others' ways as less attractive. This belief enters into why we are all different. As we become different from others, we fight for what we have, we defend the ways that we are used to. We are reluctant to become "like them," and we do what we can to protect ourselves. Who wants to be like strangers anyway? If they try to force us, we use force. If they try to convert us, we pull back. Not only do we try to maintain our differences, but conflict with others actually encourages us to hold on to our differences as much as we can.

Summary and Conclusion

You and I exist in a social context. Where we happen to live our lives and with whom we live will influence who we are, what we do, and what we believe. This, in turn, will make you and me different from each other. The intensity of our interaction, the history, problems, and patterns of our organizations, together with the feelings of ethnocentrism that inevitably arise in organizations, will keep us different, and although it might seem someday that you are becoming more and more like me (or vice versa), we should expect that differences will remain and that they will always be substantial.

It is easy to forget the role of social organization in creating the differences between people. In the 1990s we are too often attracted to racial or biological differences as explanations. It is easy to explain differences on the basis of how people look physically; it is too easy to equate belief and behavior with physical appearance. Biological differences, although they are sometimes important for understanding individual differences, seem to be much less relevant to understanding differences between groups or societies.

We can never have a world where all agree and cooperation is perfect. When it really comes down to it, why should we want that anyway? Human differences are not necessarily bad, and a very strong case can be made that they are good. Diversity, for example, encourages a dynamic approach to understanding anything in the universe. It encourages us to evaluate who we are, how we live, and what we believe. Diversity brings alternatives to what we know, new solutions to problems we encounter, and new meanings to our lives. It can teach us respect for differences and humility concerning our own views of reality. It can bring a people a much richer democracy, because it can teach them mutual respect rather than simply acceptance of what the majority wants.

And ethnocentrism? Is that good or bad? It depends on our values, of course. It seems that ethnocentrism may contribute to social solidarity and social order. It helps bind us, and it creates in us a commitment to society. It makes it easier to follow the rules, because the rules seem right. Ethnocentrism makes us feel good about who we are and more certain about what we believe. It gives us an anchor; it helps us decide what is and is not good in the world. It encourages our community to retain its unique qualities. Some ethnocentrism is undoubtedly necessary for the continuation of society.

On the other hand, ethnocentrism is costly. From the standpoint of society it discourages innovation and change as well as the solution of serious problems. People become opposed to change when it is perceived to threaten qualities in society that they cherish. There is a tendency to ignore serious social problems, because their solution is not worth giving up what we cherish. In fact, ethnocentrism discourages us from finding creative approaches to solving problems, because we are also generally committed to our particular way of solving our problems.

Ethnocentrism gets in the way of some important things that many of us value. It hinders our understanding of other people, for it makes us too quick to judge. It encourages narrow-mindedness and an unwillingness to recognize many human differences for what they are. Not only does ethnocentrism stand in the way of understanding others, but also it hampers us in understanding ourselves, because we never appreciate the fact that we, our society, and our society's rules and truths are, to a great extent, part of a social reality. Ethnocentrism tends to confuse culture with truth; it makes us feel that what we believe is true and right rather than socially developed and open to criticism.

Finally, ethnocentrism too often encourages and justifies inhumanity. It is used by political opportunists to gain support for persecution of and war against others. It is used to justify stealing from and enslaving others. It leads to persecution of minorities and destruction of individuals whose only sin is that they are different from the rest of us.

The dilemma is that ethnocentrism fosters the continuation of society and the security of its members yet undermines important qualities that most of us say we adhere to.

REFERENCES

The following works deal with human values and their development in society, with ethnocentrism, or with how differences between groups arise.

Becker, Howard S. 1973 *Outsiders*. Enlarged ed. New York: Free Press.

Berger, Peter L., and Thomas Luckmann 1966 *The Social Construction of Reality*. New York: Doubleday.

Durkheim, Emile 1893 *The Division of Labor in Society*. 1964 ed. Trans. George Simpson. New York: Free Press.

Durkheim, Emile 1915 *The Elementary Forms of Religious Life*. 1954 ed. Trans. Joseph Swain. New York: Free Press.

Eliade, Mircea 1954 *Cosmos and History*. New York: Harper and Row.

Erickson, Kai T. 1966 *Wayward Puritans: A Study in the Sociology of Deviance*. New York: John Wiley.

Erickson, Kai T. 1976 *Everything in Its Path*. New York: Simon and Schuster.

Geertz, Clifford 1965 "The Impact of the Concept of Culture on the Concept of Man." In *New Views of the Nature of Man*. Ed. John R. Platt. Chicago: University of Chicago Press.

Geertz, Clifford 1984 "Distinguished Lecture: Anti Anti-Relativism." *American Anthropologist,* 86:263–278.

Goffman, Erving 1963 *Stigma: Notes on the Management of Spoiled Identity*. Englewood Cliffs, NJ: Prentice-Hall.

Goode, Erich 1984 *Deviant Behavior: An Interactionist Approach*. 2nd ed. Englewood Cliffs, NJ: Prentice-Hall.

Herskovits, Melville Jean 1972 *Cultural Relativism*. Ed. Frances Herskovits. New York: Random House.

Hostetler, John A. 1980 *Amish Society*. Baltimore: Johns Hopkins University Press.

Jones, Ron 1981 *No Substitute for Madness*. Covelo, CA: Island Press.

Keiser, R. Lincoln 1979 *Vice Lords: Warriors of the Street*. New York: Holt, Rinehart and Winston.

Kephart, William M. 1986 *Extraordinary Groups: The Sociology of Unconventional Life-Styles*. 3rd ed. New York: St. Martin's Press.

Lofland, John 1966 *Doomsday Cult*. Englewood Cliffs, NJ: Prentice-Hall.

McCall, George J., and J. L. Simmons 1978 *Identities and Interactions*. New York: Free Press.

Rokeach, Milton 1969 *Beliefs, Attitudes, and Values*. San Francisco: Jossey-Bass.

Rubington, Earl, and Martin S. Weinberg 1987 *Deviance: The Interactionist Perspective*. 5th ed. New York: Macmillan.

Shibutani, Tamotsu 1955 "Reference Groups as Perspectives." *American Journal of Sociology, 60*:562–569.

Shibutani, Tamotsu 1970 "On the Personification of Adversaries." In *Human Nature and Collective Behavior*. Ed. Tamotsu Shibutani. Englewood Cliffs, NJ: Prentice-Hall.

Shibutani, Tamotsu 1986 *Social Processes: An Introduction to Sociology*. Berkeley: University of California Press.

Simmel, Georg 1908 "Conflict." In *Conflict and the Web of Group Affiliations*. 1955 ed. Trans. Kurt H. Wolff. New York: Free Press.

Sumner, William Graham 1906 *Folkways*. 1940 ed. Boston: Ginn and Company.

Toennies, Ferdinand 1887 *Community and Society*. 1957 ed. Trans. and ed. Charles A. Loomis. East Lansing: Michigan State University Press.

White, Leslie A. 1940 *The Science of Culture*. New York: Farrar, Straus and Giroux.

Whorf, Benjamin Lee 1956 *Language, Thought, and Reality*. New York: John Wiley.

Whyte, William Foote 1955 *Street Corner Society*. Chicago: University of Chicago Press.

Yinger, Milton J. 1982 *Countercultures: The Promise and Peril of a World Turned Upside Down*. New York: Free Press.

8

Why Is There Misery in the World?

The Consequences of Inequality, Destructive Social Conflict, Socialization, and Alienation

*I*n his book *Beyond the Chains of Illusion* (1962), Erich Fromm describes three events that inspired him to become a social scientist. The first was the suicide of a dear friend, right after the death of her father. The second was World War I, a war fought by "civilized" nations against one another, each claiming justice on its side. The third was the mass murder of the Jewish people during World War II by one of the most advanced societies in the world. These three events pushed Fromm to try to understand human beings in order to create a more just world.

Events such as suicide, genocide, and war cry out for explanation for at least two reasons: their causes are difficult to understand, and their costs in human misery beg for a solution. All of Fromm's work was an effort to understand the actions of human beings and the reasons why there is so much misery and injustice in the world. Fromm's quest is similar to those of many other great thinkers and should be important to all of us.

Sociology has always attracted scholars driven by a desire to make sense of misery and to bring justice to the world. Karl Marx, reacting to horrible conditions of poverty and the accumulation of wealth by a few people, was inspired by a vision of equality for all. Emile Durkheim, reacting to conditions of rapid social change and rising individualism, sought a world of people bound together through a shared sense of morality. American sociologists, reacting to problems of migration, urbanization, poverty, and social inequality, were inspired to create a practical science applied to serious social problems. Indeed, like Fromm, many sociologists begin their intellectual journey because of their desire to improve the human condition. Auguste Comte, the nineteenth-century founder of sociology, believed that he was founding an academic discipline that would save humanity by studying and solving the problems that plague humankind. Comte undoubtedly exaggerated what sociology could do, but there is still a faith in most of us that sociological knowledge can make a substantial contribution to improving the world.

Strangely enough, it is difficult to define misery. *Unhappiness* and *suffering* come close, but unhappiness is less acute, and both terms imply a more temporary state. Everyone is unhappy sometimes; everyone suffers occasionally. Nevertheless, many people live in a state of chronic suffering and unhappiness, which comes closer to the meaning of misery. *Misery* is used to describe certain *conditions*, such as war, poverty, and oppression. *Misery* is also used to describe how people *feel*, how they perceive their lives; in this sense misery is a *state of mind*. Both external conditions and state of mind are elements of misery, and usually—but not always—they are related. This chapter will examine both elements, but the focus will be on the conditions.

There are many approaches to understanding misery. Psychologists and psychiatrists test and treat people who are schizophrenic, paranoid, or suicidal, who lack self-worth or self-control. They have identified some very important clues to why misery exists, including chemical imbalance, genetic predisposition, early childhood training, trauma, and failures in school and friendships. Religious leaders and religious philosophers normally look to spiritual causes and call for spiritual solutions. They ask why so many of us live without religious or ethical principles.

Thoughtful religious people ask yet another and more general question about misery: How does a just God allow a world in which so much misery exists? Rabbi Harold Kushner (1981) uses the story of Job to illustrate the problem. Job is described in the Bible as a very just man. Most people consider God to be both a just God and an all-powerful God. How is it possible, Kushner asks, for an all-powerful and all-just God to curse this just man, Job, with misery?

The rabbi answers: You cannot believe all three ideas at the same time—two of them, but not all three. Either Job is not so just, God is not just, or God is not all-powerful. For example, if Job is not just, then an all-powerful, just God cursing him with misery makes some sense. Or if an all-powerful God is not just, it makes sense that this God will bring misery to a just man. After further analysis, Kushner tells us that yes, Job was a just man because the story makes no sense without this starting point. And to Kushner, God must be just, because he finds little sense in believing in a God who is not just. We are left with the third assumption: the all-powerful God. Things of this world, Kushner concludes, are not caused by an all-powerful God. They are the result of natural cause. The misery of this world is therefore not an act of God; it occurs within the natural order. Job is miserable because bad things happen to both good and bad people, Kushner explains. Disease strikes the one who catches the germ, not the one who is unjust. The earthquake destroys the property that happens to be in the vulnerable place, not the property owned by an evil person. Neither the germ nor the earthquake distinguishes between the just and the unjust. Thus, to understand disease and earthquakes we must turn to understanding nature, not God. To understand human misery we must understand natural—to a great extent, social—causes. Misery occurs because of certain conditions present in nature and society. Scientists, in general, try to discover these conditions. Sociologists, in particular, focus on those conditions that are social.

It should come as no surprise that when sociologists look at the problem of misery, they focus on *social conditions* as the cause. That is what drives our investigation. Of course, we recognize the existence of other causes, but social conditions are what we concentrate on. Durkheim cautioned long ago that if one is going to understand *social matters*, one should look to *social causes*. If poverty, violence, crime, oppression, and meaningless work are *social* matters (and they are), we must look to a *social* explanation. If suicide, drug abuse, and fear are widespread in *society* (and they are), we must examine the nature of that society to understand these seemingly individual acts. Sometimes it is the breakdown of society that is responsible; sometimes it is actually the successful operation of society that is to blame.

Sociologists look first to *social inequality* as the source of social problems underlying misery. Poverty, oppression, exploitation, and lack of hope and self-worth bring misery to many people, and these are linked to social inequality. Inequality also produces institutions—for example, public schools, private health care, and a criminal justice system that favors those who can pay—that cannot and do not work for large numbers of people, resulting in miserable conditions for many. Finally, societies built on inequality will produce people who, no matter what they have, feel misery even if they do not live in obviously deprived conditions.

Widespread *destructive social conflict* and the breakdown of social order is the second source of human misery, from a sociological perspective. Society to a great extent is a cooperative order. It is built on trust and agreement. Conflict is necessary in and contributes to society, but conflict sometimes becomes destructive, disrupts and destroys life, and creates chaos. For many people, it brings fear and a feeling of vulnerability to the whims of others.

Third, sociologists focus on *socialization*. Human beings are socialized, or taught the ways of society, from birth to death. In many complex ways, socialization creates misery for people. For some, socialization is inadequate, and the individual lacks proper social and emotional support or does not learn the self-control needed for successful problem solving. For others, socialization teaches moral rules encouraging exploitation and destruction of others. For still others, socialization creates unrealistic expectations, so that no matter how successful one is, one cannot overcome a feeling of misery.

Finally, several important traditions in sociology look to *alienation* as a cause of misery. Alienation is the separation of people from one another, from meaningful work, and from one's self (that is, a sense of ownership over our most fundamental possession, our self). Conditions in society create alienation of the individual, and this alienation is a fourth cause of human misery.

The accompanying figure summarizes these four general causes of misery. Each cause will be discussed in more depth in the rest of the chapter.

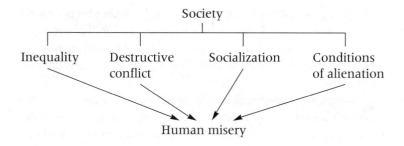

The First Cause of Misery: Social Inequality

Inequality brings misery to many people. In the struggle to succeed, they (or their predecessors) lose out and end up dependent, power-less, or exploited. Their lives are without hope, and they often strug-gle simply to survive. They turn their anger inward or toward others, who, in turn, become victims. Inequality may also encourage those who succeed to exploit and destroy others to get more.

Without question, there are many justifications for a society built on the principle of inequality. We often hear this (expressed in the promise of material rewards) from political leaders, media, pro-fessors, and family: "Work hard, and you, too, can achieve a life of privilege. You, too, can rise above all the rest. You, too, can have material success." "Competition brings out the best in the human being, and without it we would not succeed as individuals or as a society." "Promising great rewards to those who work hard and act smart is basic to building the good society." "In order to encourage people to take chances and in order to get them to take difficult and responsible jobs, we must give them hope for great material re-wards." "People have a moral right to keep whatever they make." In virtually every society where inequality exists, there is a philosophy that attempts to justify it.

Inequality, by definition, means that in social relationships, from dyads to societies, some people are favored over others. Where there is inequality, some will be high on the social ladder, and some will be low. All cannot rise to the top. All cannot receive an "A" in a class based on the normal curve. All cannot be millionaires in a cap-italist society. Indeed, one of the main reasons there are millionaires in the first place is that some people are able to get others to work for them at a rate far lower than what they (the millionaires) receive. Some receive a little less (the upper middle class); most receive quite a bit less (the working class); and many receive almost nothing (the poor).

When we relate inequality to social class, we must note two important aspects. First, in a class society, one is advantaged or disadvantaged based on what economic resources one accumulates. And second, one tends to pass down advantages or disadvantages to children through educational opportunities, social contacts, or direct inheritance. Inequality is perpetuated, and where people are distributed is perpetuated. The net effect is that over time great inequalities arise in the distribution of resources within society and among societies in the world. Some people live in splendor; some are affluent and secure; many can barely pay their bills; and many live on the verge of starvation. A sensitive tourist to any big city in the United States cannot help but see people without food, adequate clothing, shelter, or dignity crying out for help while others casually walk into art galleries and pay $100,000 for a painting.

Consequences of Inequality: Poverty

The critical question is: What happens to those left behind in society? From 15 percent to 20 percent of the American population lives in poverty. Poverty is associated with a range of disadvantages that profoundly affect how people lead their lives: lack of productive work, unemployment, poor physical and mental health, family stress and disorganization, lack of educational opportunity, inadequate protection under the law, and drug and alcohol abuse. Poverty strains people's ability to solve everyday problems; any planning for the future must be sacrificed to everyday survival. Many of our most serious social problems can be traced to poverty, not only because the poor are deprived of what everyone else enjoys, but also because victims of poverty often fall prey to anger, crime, violent conflict, family disorganization, and political instability.

It is tempting to look down on the poor from more privileged heights and complain that "it's their fault" and that "we must take care of ourselves." Such attitudes help to perpetuate the unequal distribution of resources; redistribution becomes a major challenge that most of us are not motivated to take on. The fact is that poverty is built into a society of inequality. It results from a system in which some people make it at the expense of others, a system in which some people are born into situations in which opportunities are fixed against them, a system in which social change favors some and leaves others behind.

Consequences of Inequality: Crime

In a society of great inequality, people are socialized to judge themselves and others on the basis of material success: "I am good because I have achieved much in this competitive game of life." "Others have more than I, but maybe I can get there, too. Others have less, and I'm fortunate." Material success is a value many people share. A society of inequality bestows dignity on those who rise to the top and withholds dignity from those who remain below.

The game of life is fixed, however. Opportunities are never equal. We are all born into privilege or lack of it; class is largely an inherited rank. We learn what success means in society, and the choice becomes clearer and clearer to those in inferior positions: either accept a lowly position or work to change your position. And if you work to change your position, there is another choice: work extra hard in a system that favors others, or go outside the legitimate system to make it. Many poor people accept their position and struggle simply to survive. Many work extra hard to make it in the legitimate order. But others see no reason to follow laws that seem to work against them in the competitive order, laws made by those who most benefit from that order. Stealing, prostitution, selling illegal drugs, and violent crime become attractive options. Some will overcome poverty through crime; the vast majority will not. Those who do not will remain poor, increasingly victimized by the welfare, court, medical, and prison systems that attempt to exercise control over their lives to ensure that they are not threats to the rest of society. Over time their misery worsens.

The poor are by no means the only ones who break the law and try to achieve success outside legitimate means. Crime exists at all levels of society because of the widespread inequality and the passion of people to improve their rank. It exists because the rich try to stay rich or get richer. And although those who succeed in improving their rank illegally may or may not overcome their misery, those who are caught and punished will have to deal with additional problems. Politicians who take bribes, stock brokers who deal illegally, and employers who do not protect employees from hazardous waste are all examples. Almost always, however, their situation will not come close to the misery of the unsuccessful lawbreaker who is among society's poor. That is because the court system gives harsher treatment to the poor; the wealthy are more able to escape prison through paying fines, hiring expensive lawyers, and convincing the courts that they are not a danger to society.

When we think of crime, most of the time it is not the perpetrators but the victims whom we think of as living lives of misery. Those who are the victims, however—those who are preyed on by the lawbreaker—are also disproportionately the poor. They are the ones whose neighborhoods are infested with gangs and drugs. Organized crime infiltrates poor neighborhoods through providing illegal goods (handguns, stolen goods) and services (prostitution, fencing). The poor are in close proximity to those who engage in crime, they are the ones who are most open to exploitation, and they are the ones least likely to be protected by the system of law.

The point should not be missed. In large part the cause of crime in society is inequality. It results from the fact that almost everything in society teaches us that being more successful materially is what makes life worthwhile, causing some to see crime as the easiest way to achieve that success. Price fixing, drug dealing, and bank robbery are some of the consequences of a society that emphasizes material success.

Consequences of Inequality: Bad Jobs

But inequality fosters more than poverty and crime. It generates tedious, low-paying, dangerous, and insecure work for many. The work that many people do offers few material rewards; it traps them in a life of bare *survival*. Misery exists, in part, because miserable work exists; those who have little choice in the matter must take it or die. Consider mining, for example. For many generations, people in Appalachia have taken low-paying, tedious, physically demanding, dangerous, and insecure jobs as miners. Why? Because "someone has to do them" and because those who own the mines can profit only if those who work for them remain poor. If the workers become materially successful and expect decent wages, the rich will find that mines are no longer profitable and will have to close them down. So the poor must choose between no work and bad work. The same is often true for the women who clean our homes or who care for our children: their job security depends on low wages. In other words, it is their "willingness" to take low-paying jobs that guarantees their continued work. Bad jobs and low-paying jobs will always be a part of a society in which some succeed at the expense of those who need these jobs to survive.

Bad jobs are also the most insecure. Those at the bottom of the employment ladder are in unskilled occupations: those most likely to be replaced by machines, by labor in other societies, or simply by other workers willing to work for less. In times of depression, their jobs are the first to go, and they are the most likely to experience long-term unemployment.

Work is a major part of a human being's life. Miserable work contributes significantly to a miserable existence.

Consequences of Inequality: Exploitation

Inequality, Marx believed, causes misery in yet another way: it can always be translated into power (Marx and Engels, 1848). Where inequality of any kind exists, unequal power is almost inevitable. It does not matter if that inequality is based on economic resources or political, occupational, gender, racial, or religious position in society. Where power is unequal, exploitation (selfish use by others) is made easy, because those with power are in the position of demanding things from those who fear or depend on them. Marx emphasizes economic exploitation: those who own the means of production (factories, for example) are extremely powerful, so they are able to exploit all who must depend on them for work.

Sociologists have gone beyond Marx in their analysis of inequality and power, however. Work is not the only basis for power, dependence, and exploitation. Abused wives or children continue to be abused, in part, because the physical inequality is so great, because they are made relatively powerless through relying on the man's paycheck, and because they fear the consequences of challenging or leaving him. And in many societies in the world as well as in many communities in the United States, governments do not adequately protect children and women from abuse. The dominant man controls, threatens, and exploits the dependent woman or child. Nonwhites are exploited by whites, the defenseless by the violent, the small business by the large corporation. Inequality of all kinds means dependence, and dependence facilitates exploitation, which leads to misery.

Exploitation of the powerless characterizes almost every society. Our own history, which we too often idealize, has been one in which African Americans were enslaved; Asian Americans, immigrants from Southern and Eastern Europe, and Mexican Americans were used as cheap labor; and Native Americans were victims of our desire for good land. Most European societies discriminated against and exploited Jewish people, and most persecuted and exploited Christians who were not part of the dominant denomination. Fear, anger, physical expulsion, execution, extermination, and denial of rights and privileges enjoyed by the dominant group are but some of the instances of misery brought on by such systems of inequality. And, of course, where such inequality still exists, exploitation and misery continue.

Almost every society has also had a system of inequality based on gender. Where gender inequality is extreme, sexual exploitation is regarded as normal and legitimate, the destruction of infants because they are female is an accepted practice, and the physical abuse of wives by husbands is a right. Gender inequality closes off women from equal participation in the political, economic, and social orders. It denies educational opportunities and legal rights that men enjoy. Men enjoy privilege at the expense of those in a less powerful position.

Consequences of Inequality: Lack of Self-Worth

All forms of inequality—economic, racial, ethnic, religious, and gender—bring misery to those in lower positions simply because those people are looked down on. The poor are defined as undeserving and lazy; nonwhites are seen as less capable than whites; and women are defined as submissive, passive, inferior intellectually, and as sex objects for men. It is clear that these beliefs are used to justify the system of inequality and the discriminating acts against the exploited groups. It is also clear that these beliefs tend to "objectify" the dispossessed. Those people may be human, but they are somehow different from the rest of us, objects rather than real human beings with feelings, and this status makes their misery easier for us to accept.

Of course, such beliefs also contribute to the misery of those who are in the dominated groups, because many come to believe what they are taught about themselves. Misery exists, in part, because people disparage themselves; they disparage themselves, in large part, because others define them as less worthy and because they see that those who are similar to them are also defined in that way. If those in powerful positions are able to justify their status by claiming that they "worked hard" or that they "are superior," "smarter," or "more talented," what does that imply about those who are in lower positions? That is part of the misery of the oppressed: being told and then coming to believe that they are somehow to blame for their position or that their gender, color, or religion automatically makes them less worthy.

Great miseries result from lack of self-worth: mental illness, alcoholism, drug abuse, and suicide. Although all such conditions cannot be traced to the struggle for material gain in a society that measures dignity by one's level of material success, many can. Misery takes a toll, and for some in society the toll is great. In every one of these problems—lack of self-worth, mental illness, alcoholism, drug abuse, and suicide—it is the poor who suffer the most, but all who think they fail to measure up are vulnerable.

Consequences of Inequality: Stress

Social inequality also teaches us what is "important" in life. It creates expectations in all of us. A society of inequality presents us all with a race to be run, a "rat race," as some would call it. We see how those above us live, and those below us. We see that human dignity depends on where we are in this game. The central ideas, values, and rules of culture create in most of us a commitment to winning and achieving material success. And what this arouses in most people is *fear*—of losing what they have—and *hope*—of rising above where they are. Both fear and hope stimulate effort, but fear that is realized or hope that is continuously frustrated brings misery. Misery can come to those at the top as well as to those on the bottom, because it arises in part from one's perceived lack of success in the system of inequality. Of course, the misery of the poor is compounded by the ever-present problem of bare physical survival.

The fight to stay even or do better brings with it the temptation to commit crime, and this, in turn, may bring more misery both to those who commit crimes and to those who are the victims. The simplest illustration is drugs. Crime associated with drugs is the work of both the poor and the wealthy; all are interested in improving rank in society. Again, one must not forget that the poor are always the most likely victims, who tend to be exploited by the more rich and powerful.

The study of the homeless should remind all of us of a simple truth: many of us are on the verge of homelessness. We go to school, get a job, work hard, buy our home, pay the mortgage, and hope to live happily ever after. However, we can suddenly find ourselves thrust out of work or can go broke in our businesses. Our companies and communities can close up overnight, and house values can plummet as international economic forces play themselves out. Or we can marry, have children, depend on a spouse, and suddenly find ourselves out of a marriage and into poverty. Or we can retire and suddenly find it impossible to survive on the little we have accumulated.

Consequences of Inequality: Institutions that Produce and Maintain Misery

Finally, to understand how inequality produces misery in society, it is important to understand what *social institutions* are. Institutions are the various practices in society that are supposed to meet its ongoing needs and alleviate its problems. For example, we have established political institutions to determine goals and organize our efforts to achieve those goals; legal institutions to protect social order; educational institutions to socialize people into society and help develop their abilities; religious institutions to hold us together, make morality sacred, and help give life meaning; and economic institutions to produce, distribute, and consume goods and services. In a society of inequality, however, such institutions do not and cannot work for all. Institutions are created over time as society develops; they are created through a competition among people who want them to serve them. The inequality of power in society plays itself out in that competition, so it is the more powerful who have more say in what those institutions are and how they work. Of course, they normally work best for the powerful who created them, control them, and run them. Thus, *they are not usually meant to solve problems of human misery*, except when misery touches the lives of those who are powerful.

Our system of private medicine and private health insurance primarily takes care of the needs of those who can afford it. Public and private education, supposedly set up for the purpose of helping people raise themselves up in the system of inequality, normally function to keep them approximately where they began. Our system of law, our political party system, and our court and prison system protect and benefit primarily those high in the class system and, at the very least, function to keep stable the system of inequality that prevails in society. For fifty years, Eastern European societies built political, legal, economic, and educational systems that clearly came to benefit the politically powerful at the expense of everyone else. Throughout American history, we can identify a disproportionate number of institutions that benefit the wealthy and middle classes at the expense of the poor and working classes, whites at the expense of nonwhites, and men at the expense of women. Our efforts to correct misery in society have never equaled our commitment to a society built on the principle of inequality, so our institutions have changed slowly, usually when less powerful groups have organized and demanded that the powerful make changes.

The Second Cause of Misery: Destructive Social Conflict

The Meaning of Destructive Conflict

We come now to the second cause of misery in the world: destructive social conflict.

Not all conflict is destructive. Indeed, it is important to see most conflict as inevitable, necessary, and constructive. Conflict means that human beings in social interaction struggle with one another over something that they value but cannot all achieve. Conflict is interaction in which actors use power—try to impose their will—in relation to one another. Competition is one form of conflict: it is conflict that takes place within clearly specified rules. Whenever actors try to persuade one another, whenever they fight one another for a cause they believe in, conflict arises. Whenever we try to achieve our goals and others are involved, there will invariably be some struggle, and usually negotiation and compromise result. The result of conflict is usually positive: both parties get something, organizations change, people's interests are heard, and problems are identified and dealt with. Constructive conflict is a fact of life, and instead of causing misery, it is one way in which misery can be recognized and alleviated.

Destructive conflict is something else. Wars are fought, and people are killed or made homeless, their lives left in ruin. Riots cause physical harm, killing, and the destruction of property by both rioters and the authorities. Spouse and child abuse physically harm people in the short run and cause destructive emotional effects in the long run. There are always victims in destructive conflict.

Destructive conflict is characterized by intense anger and the desire to destroy or hurt one's opponent. Such conflict often escalates and becomes increasingly violent, inflicting physical and emotional harm on the victims. It ends in harm to others while ignoring the real issues between people.

The Causes of Destructive Conflict

Why does conflict become destructive? Why does something that has every potential for contributing to human welfare become a source of human misery?

For one thing, constructive conflict is often ignored, and thus differences are neither faced nor resolved. The more powerful refuse to recognize its existence. The less powerful fear to express interests that might bring conflict out in the open. Or conflict sometimes seems irreconcilable to the parties involved, and thus there seems to be little point in trying to resolve it constructively (this used to be the case between Israelis and Palestinians). Often, people run from constructive conflict out of fear that it might escalate into highly aggressive and even violent confrontation (family conflict is an example). If conflict is repressed over time, however, it grows more intense and emotional. The goals that each party was originally seeking are lost, replaced by hostility rather than goal-directed efforts to resolve differences.

Social inequality is an important source of destructive conflict. Violent revolutions arise from inequality, often begun by people who are rising in the social order yet still feeling left out. Violent crime often arises from the frustration and anger fostered by inequality. Wars are often a result of one nation attacking another because it has superior resources and wants something that the other has. Aggression often occurs because inequalities within society are not faced, problems are externalized, and leaders try to bring unity within the nation by creating a common enemy. Individuals who are deprived in a system of inequality become frustrated, angry, and violent offenders: against family members, against strangers, against the successful, and ultimately against themselves.

Destructive conflict also occurs because many of us have learned to use violence to deal with problems we face. In trying to achieve their will in relation to others, people learn to use violent confrontation. American society is more violent than some societies and less violent than others. Our political leaders through what they say and do tell us that it is all right to resolve problems through violent, destructive conflict. In how parents act toward children, they, too, express this message, and movies, television, and even music reinforce it. Cartoon characters, superheroes, and men who must prove their manhood through aggressive violence are important examples to us. Several themes in our history teach us that destructive conflict is necessary and even good: the winning of the frontier, Western vigilantism, wars of expansion, slavery, and violent oppression of minorities are examples. Of course, there are also values, principles, and institutions in American society that limit violent destructive conflict: participating in the democratic process through voting, a spirit of compromise and negotiation in politics, a reliance on law, and usually a respect for individual rights.

To the extent that culture encourages individuals and groups to use violence, it is encouraging destructive conflict and human misery. Whenever government, family, or other leaders legitimate the use of violence, they are displaying to others that violence is one way in which problems can be solved.

Misery as a Consequence of Destructive Conflict

Destructive conflict hurts the victim of violence. It does not matter where it comes from: parents, police, lawbreakers, labor unions, management, or government. It is meant to hurt or destroy the other, and it often does. It produces anger in those who are victims. It may bring quiet anger, and often that anger becomes chronic and deep-seated. It may be expressed, or it may simply fester. Though it is often aimed at others, it can also be aimed at oneself in the form of self-destructive behavior: crime, drug or alcohol abuse, or suicide.

There is every reason to believe that even if one wins in destructive conflict, misery is far from being eliminated. The cycle of destructive conflict is difficult to halt. Winning brings anger on the other side and the possibility of retaliation in the future. Winning causes one (and others) to believe that destructive conflict is the way to achieve what one wants in situations, thus encouraging its continued use. Psychologically, it encourages more—rather than fewer—aggressive feelings in the perpetrator. Instead of making one feel good, aggression tends to make one more angry and destructive toward the victim. That is because aggressors justify violence through dehumanizing the victim, convincing others as well as themselves that "the victim deserved it." Such a belief system encourages further aggression.

Violence and destructive conflict often occur outside the legitimate order. They are not what we expect from one another. They are actions that flout convention. Social interaction depends on convention and the underlying idea that those around us—even strangers—will follow that convention. Interaction thrives on trust, and one of the real victims of destructive conflict is trust. Through violent conflict the world of the predictable and familiar becomes a world of disorder and unpredictability, without rules for people to put their faith in. A world of distrust, rule breaking, and unpredictability makes life miserable for many. They become victims of the strongest; they become afraid of the world that used to be taken for granted.

Misery in society can be traced to many root causes. We have thus far focused on two: social inequality and destructive conflict. A theme weaves itself through both of these causes: that which seems to be an integral part of society, even necessary for its continuation, also brings misery to many people within society. Inequality, so much a part of our society—and even what some would call a strength—brings misery to large segments of the population, and that misery, in turn, can bring misery to the rich and powerful and threaten the continuation of social institutions that most people have come to take for granted. Social conflict, a necessary part of all societies, becomes violent and destructive, and destructive conflict destroys victims, harms perpetrators, and undermines society itself.

The Third Cause of Misery: Socialization

All sociologists recognize the significance of socialization on the kind of individuals we become. Socialization influences our choices, teaches us the rules by which we control ourselves, and finishes what nature has begun by forming our most central qualities as human beings.

Some people are born into a world in which socialization is insufficient or inadequate. Early interaction within families is sometimes too limited or destructive or fails to express love and teach self-control. Close relationships are vital to human beings. We need affection as children if we are to become fully human. Positive emotional ties provide us with the raw materials for intellectual and emotional growth. There is strong evidence that infants without close ties to others die or are psychologically harmed. There is also evidence that deprivation of affection in the early years of development has serious emotional and behavioral consequences later on.

People who are not loved do not usually love themselves. People who do not have close ties in childhood have a difficult time developing close ties later on. A life without affection in the early years of socialization is an important source of human misery, both for the actor and, often, for others with whom the actor interacts.

Besides the emotional support and close ties that early socialization must provide, it is also important for the development of *self-control*. We learn the ways of society, and those ways become ours; we internalize them, and we control ourselves accordingly. Some of us do not learn to control ourselves very well. Our conscience is weak, or the drive to ignore it is strong. Two serious problems arise: (1) others become victims of our lack of control, and we create misery for them; and (2) we have problems in cooperating with others, making it difficult to meet our own needs successfully.

Socialization also has a great impact on the choices we make and the direction we follow in our lives. It influences our view and use of illegal drugs, the value we give to our education, our choice of major and occupation, our commitment to the law or rejection of it, whether we marry, whom we marry, and how we treat our spouse. It matters *who* socializes us. Our parents, teachers, adults in our neighborhoods (in businesses, churches, and on the street), and friends influence our directions. We observe them, listen to them, and watch how they react to our ideas and acts, and through it all we try directions that seem to be right for us. It matters what we are taught by people with whom we interact. In a sense, they represent society to us, and we are influenced by their rules, their values, their ideas, and their example. In *Tally's Corner* (1967), a study of a lower-class ghetto community in Washington, D.C., Elliot Liebow dramatically shows us that the younger men there learn how they should act by observing how the older men behave. The older men spend their time hanging out. They work at temporary, low-paying, un-skilled, and often dangerous jobs. Jobs do not offer them hope; they offer only an opportunity to get through the week. With little hope for the future, these men take whatever pleasure they can get in the present. Life is bare survival, with little dignity except what one can get from the others on the street corner. The younger men come to believe that these older men are what they can expect to become; in these others they see their future selves.

Of course, we are also socialized by people we never personally meet. We might be influenced to change our directions by a political leader, a basketball star, a successful singer, a wealthy businessperson, or even a ruthless criminal. We may read a book or by chance interact with someone who touches us, and it is possible for our direction to change. However, it is tempting to overemphasize the importance of distant figures. Most of the time we are socialized by people who are much closer to us and with whom we interact every day. The late Supreme Court Justice Thurgood Marshall (1979:1) wrote about how difficult it was for him to go into a ghetto neighborhood pretending to be a role model for African Americans there: Their lives are a long way from his. Their opportunities are not the same as he had. They know the gulf between him and themselves too well.

Those who are socialized by people who live lives of misery are influenced to go in directions that will bring them misery. That is the reality of socialization. Through others close to us we become aware of what our lives will be and should be. We learn what we have a right to expect out of life: dropping out of school or getting a graduate degree from a leading university, barely surviving in poverty or living in affluence, getting a break or planning for a career. For many, role models use drugs, commit crimes, engage in destructive conflict, and treat other people with contempt. For others, role models are people who prey on the poor and seek material success at any cost. People live in misery, in part, because that is the direction in which their socialization takes them. To overcome the misery that one is born into, one must be socialized by realistic role models who work against that socialization and alter the direction in which one decides to go. If one is located in the midst of misery, it is almost impossible to find realistic role models to help one escape; here lies the viciousness of misery for those caught up in it.

The final way in which socialization produces misery for people involves the power of expectations. Most of us know people who "seem to have everything" yet are still dissatisfied with their lives. People who are beautiful think of themselves as ugly; people who are rich think of themselves as poor. People who get "A's" get one "B" and fall apart. Some of our misery stems from the difference between objective reality (how we actually perform, what we actually have) and our expectations for ourselves. Our expectations come primarily from our socialization. As we are socialized by others, their expectations become important to us; their demands become our own. Sometimes we rebel against overdemanding parents; more commonly we never do escape their expectations, which we can never satisfy. It is critical to face the fact that part of personal misery is social: no matter how much we get from life, we cannot be satisfied, because others have socialized us to be impossible taskmasters over what we do.

Socialization is really the link between society as it exists out there and the individual. It is absolutely essential for the continuation of society and for the development of the human being. It creates order; it creates the opportunity for the individual to achieve his or her potential. It can also create disorder in society and misery in the individual.

It is important to keep in mind the many ways in which socialization enters into the problem of misery. As we see the horrible misery that serial killers, terrorists, business tycoons, youth gangs, and drug pushers bring to others, it behooves all of us to ask *why*. Why do such people exist? If we look closely, we can almost always identify socialization as one of the most important causes. This does not excuse the harm, but it does help us see the misery that such people experience themselves, and it allows us to understand the causes of their actions and recognize how lucky most of the rest of us are.

The Fourth Cause of Misery: Alienation

Another source of misery, from a sociological perspective, is alienation. In its simplest sense the word *alienation* means *separation*. It is a concept sociologists use to describe separation *from the other people*, separation *from meaningful work*, and separation *from ourselves as active beings*.

Alienation from One Another

Alienation is a central theme in the work of Marx. Capitalism, according to Marx, is an economic order based on competition rather than cooperation, exploitation of others rather than sharing, and materialism rather than love and respect. Marx (1844) describes how, in his view, people relate to one another in capitalist societies: as things, as commodities to be bought and sold in the labor market-place, as property, and as means to an end rather than as ends in themselves.

Many other sociologists acknowledge the social alienation that modern life has brought but do not lay the blame on capitalism. Max Weber (1904–1905:181–183), although he describes the many benefits of bureaucracy, constantly reminds us that bureaucratic society is an impersonal society, one without feeling and tradition, one that emphasizes efficiency and effectiveness in organization. We are all caught within the "iron cage of bureaucracy," planning, calculating, and solving problems as they arise yet sacrificing friendship, close emotional commitments toward one another, and a sense of community. Charles Cooley (1909), an American sociologist who wrote early in the twentieth century, describes the importance of primary groups (face-to-face groups that entail close emotional ties) to the human being. He and other sociologists bemoan the fact that our world has increasingly become impersonal, associational, and individualistic. Intimacy and caring are increasingly replaced by social alienation.

Social alienation is probably best described in the work of Georg Simmel, a German sociologist who was a contemporary of Weber and Durkheim (all three died between 1918 and 1920). Simmel (1902–1903) sees modern life as the life of the stranger. We live in large communities in which our primary concern is with our personal needs, and our ties with others are without much depth. For many of us, urban life is a world of strangers, and the closeness that used to characterize human relationships is lost. The result for people in modern society is loneliness and misery.

Individuality is one of the dominant themes of our century. Revolutions have been fought to free the individual from the bonds of dictatorship. Education tends to make us more individualistic, less traditional, and less communal. The city cuts our ties with the tyranny of small-town control, and affluence brings the opportunity to pull back into our homes and enjoy life without having to interact with others. Individuality has exacted a cost, however: for many of us it has brought a separation from others, a decline in family and friendships, and a concern about self without a concern for the community within which we live. Together with the impersonality and selfish exploitation that society encourages, it has contributed to our alienation from one another.

Alienation from Meaningful Work

Social alienation in modern life is accompanied by alienation from creative work. Marx's ideas (1848) continue to be most important here. To him, the human being is a creative, hardworking, productive being. But for much of modern human history, work has meant laboring for the material benefit of owners, for wages they are willing to pay, for extrinsic rewards rather than for the intrinsic benefits found in creative work. We labor for others, and our labor amounts to contributing one small task that eventually produces a finished product that we never see. Work has lost its meaning for human beings, and this loss, too, has brought us misery.

Weber (1904–1905) sees early capitalism as a time when people did, in fact, go out and creatively build businesses that they cared about. Early capitalism was the period of the entrepreneur, the builder of goods and business, the creative adventurer who found real meaning in work. All that creativity has passed away in the huge, modern bureaucratic enterprises created in the name of efficiency. The actor is now a cog in a great machine, finding little meaning in work, accepting the security of position rather than the adventure of work.

Marx and Weber are the founders of the sociology of work. They ask some provocative questions, all concerned with the possibility of meaningful work, and conclude with an indictment of modern life as a place where humans are not able to find it easily. Making money has replaced meaningful work as a goal for most of us, and pursuing a satisfying life through leisure rather than through work has increasingly become the norm. Life for many is a struggle to win in a game that alienates us both from one another and from meaningful work.

Alienation from Our Active Self

To be alienated from oneself as an active being simply means that humans become passive in relation to their world. They give up. They allow government to rule them, employers to hire and fire them, neighbors to bother them, their children to demand and receive from them, and social forces to manipulate them. Their lives are not their own but, instead, are moved by impersonal forces that seem to be outside of their control. Passivity brings misery to many: they become victims of the whims of others, they are unable to deal effectively with problems as they arise, and, probably most important, they *feel powerless* in their personal life and in society.

The sociologist asks again and again: What is there about our social life—our society—that creates passivity and the feeling of powerlessness? In part it is an objective condition. We call ourselves a democracy; yet it is obvious that one vote matters little in a society so large, complex, and difficult to understand. We call ourselves a capitalist society; yet the market is clearly controlled by very large corporations and even larger, impersonal economic forces. We call ourselves a society in which the individual matters; yet things seem to change in directions that we as individuals do not wish. Even within our own personal lives, powerlessness is encouraged by the nature of society: the efforts of our parents are frustrated by the influence of peers, television, and the general youth culture. The choice of job and neighborhood is dictated by market and interest rates. The chances for getting a decent degree from a decent university are dictated by university regulations, the assignment of an adviser, and evaluations by admissions officers, instructors, and university administrators whom we never meet. It is conditions such as these that breed a feeling of powerlessness, and such a feeling brings misery to many people in modern society.

Summary and Conclusion

Human misery and its causes are difficult to understand. There are many many causes, some of which are not discussed here. In truth, the psychologist, the philosopher, and the religious thinker have much to say about why misery exists in the world.

The sociologist, however, tells us all something very valuable: the cause of human misery, in part, is in the nature of our society and our social life. People harm other people for social reasons that we are able to identify. People live miserable existences not because of rational free choices they make as much as because of social forces they are often not aware of or do not understand. Our social life is critical to what we all become, to whether our lives are fulfilling and productive or miserable and destructive.

It might be useful to bring together into a picture the points made in this chapter concerning the four social bases of human misery (see figure on page 195).

Can we alter these four broad social conditions and thus have an impact on human misery? Is human misery inevitable? Should we simply accept it? What difficult questions! All religious philosophies seek to understand them, as do all people who seek justice. Revolutions are unleashed by these questions, and even those who contribute to the misery of others will rationalize their inhumanity by declaring, "If we don't exploit these people, there will inevitably be others who do."

Human misery is probably inevitable, but it never has to be as great as it is. We can always create a society of *less misery*, or we can actually create a society of *more misery*. Poverty, for example, is more widespread today than it was in the 1960s; it is far less widespread than it was in the 1850s. Work is less exploitative in society than it was before the advent of labor unions and modern technology. Disease and hunger are less prevalent in the United States than in most other societies, yet some societies are far more successful than we are. Misery will continue to exist, but the question is always, *How much can I (or society) accept?*

It is important to recognize that much of the misery in the world is built into the nature of society itself. We cannot make progress against human misery without changing the social patterns that have developed over a long period. We cannot deal effectively with anger, alienation, and violent crime without lessening poverty, the extremes of inequality, and the linking of human dignity to material success. We cannot deal with alienation without asking important questions about the nature of work and the importance we give to individualism in this society. Of course, some of us do not

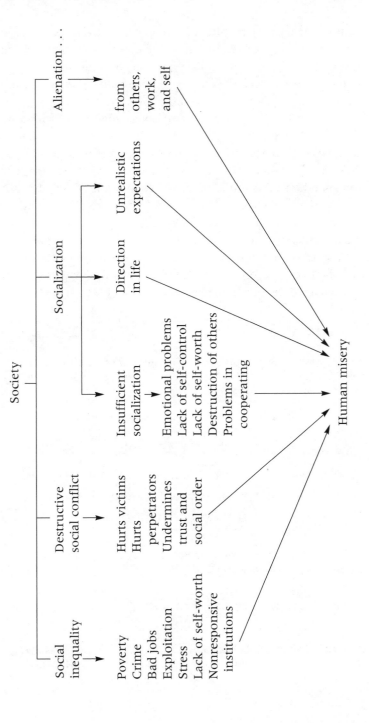

want to change society (after all, we live good lives), but then we must be prepared to accept the misery of others. If we do not change the conditions that lead to misery, it will continue and may even become worse. In fact, the misery can eventually give rise to much greater change than most of us would want.

An interesting article in *Parade* magazine asks a provocative question: "What if those children killed by abusing parents had actually lived?" Their deaths were monstrous, but what would their lives have been like? The author makes the point that those who prey on others grow out of miserable conditions. Those who bring misery to others have themselves endured miserable conditions. Being abused encourages abuse. We may not really know how to end the misery created in society, but it is folly to believe that it will go away if we simply wish it away. And it is equally folly to believe that the lives of those of us relatively free of misery will not be touched by those for whom misery is a way of life.

REFERENCES

Most of the following works deal with the consequences of inequality. Some are good introductions to crime and to alienation. All are attempts to understand human problems in society.

Adam, Barry B. 1978 *The Survival of Domination: Inferiorization and Everyday Life*. New York: Elsevier.

Auletta, Ken 1982 *The Underclass*. New York: McGraw-Hill.

Baldwin, James 1963 *The Fire Next Time*. New York: Dial Press.

Beeghley, Leonard 1983 *Living Poorly in America*. New York: Praeger.

Beeghley, Leonard 1989 *The Structure of Social Stratification in the United States*. Boston: Allyn and Bacon.

Bensman, David, and Roberta Lynch 1987 *Rusted Dreams: Hard Times in a Steel Community*. New York: McGraw-Hill.

Blauner, Robert 1964 *Alienation and Freedom: The Factory Worker and His Industry*. Chicago: University of Chicago Press.

Box, Steven 1984 *Power, Crime, and Mystification*. London: Tavistock.

Braverman, Harry 1974 *Labor and Monopoly Capital: The Degradation of Work in the Twentieth Century*. New York: Monthly Review Press.

Carmichael, Stokely, and Charles V. Hamilton 1967 *Black Power*. New York: Random House.

Chambliss, William J., and Robert Seidman 1982 *Law, Order, and Power*. 2nd ed. Reading, MA.: Addison-Wesley.

Clark, Kenneth B. 1965 *Dark Ghetto*. New York: Harper and Row.

Cooley, Charles Horton 1909 *Social Organization*. 1962 ed. New York: Schocken Books.

Currie, Elliott, and Jerome H. Skolnick 1988 *America's Problems: Social Issues and Public Policy*. 2nd ed. Glenview, IL: Scott, Foresman.

Denzin, Norman K. 1984 "Toward a Phenomenology of Domestic Family Violence." *American Journal of Sociology, 90*:483–513.

Durkheim, Emile 1893 *The Division of Labor in Society*. 1964 ed. Trans. George Simpson. New York: Free Press.

Durkheim, Emile 1897 *Suicide*. 1951 ed. Trans. and ed. John A. Spaulding and George Simpson. New York: Free Press.

Eliade, Mircea 1954 *Cosmos and History*. New York: Harper and Row.

Erickson, Kai T. 1976 *Everything in Its Path*. New York: Simon and Schuster.

Erickson, Kai T. 1986 "On Work and Alienation." *American Sociological Review, 51*:1–8.

Eshleman, J. Ross 1991 *The Family: An Introduction*. 5th ed. Boston: Allyn and Bacon.

Ewen, Stuart 1976 *Captains of Consciousness*. New York: McGraw-Hill.

Fanon, Frantz 1963 *The Wretched of the Earth*. New York: Grove Press.

Farley, John E. 1987 *American Social Problems: An Institutional Analysis*. Englewood Cliffs, NJ: Prentice-Hall.

Farley, John E. 1988 *Majority–Minority Relations*. 2nd ed. Englewood Cliffs, NJ: Prentice-Hall.

Feagin, Joe R. 1975 *Subordinating the Poor: Welfare and American Beliefs*. Englewood Cliffs, NJ: Prentice-Hall.

Feagin, Joe R., and Clairece Booher Feagin 1986 *Discrimination American Style: Institutional Racism and Sexism*. 2nd ed. Malabar, FL: Kreiger.

Finestone, Harold 1976 *Victims of Change*. Westport, CT: Greenwood Press.

Fromm, Erich 1962 *Beyond the Chains of Illusion*. New York: Simon and Schuster.

Glasgow, Douglas G. 1980 *The Black Underclass*. San Francisco: Jossey-Bass.

Goffman, Erving 1961 *Asylums*. Chicago: Aldine-Atherton.

Goffman, Erving 1963 *Stigma: Notes on the Management of Spoiled Identity*. Englewood Cliffs, NJ: Prentice-Hall.

Gordon, Milton M. 1978 *Human Nature, Class, and Ethnicity*. New York: Oxford University Press.

Harrington, Michael 1963 *The Other America*. New York: Penguin Books.

Harrington, Michael 1984 *The New American Poverty*. New York: Penguin Books.

Henslin, James M. 1990 *Social Problems*. 2nd ed. Englewood Cliffs, NJ: Prentice-Hall.

Horton, Paul B., Gerald R. Leslie, and Richard F. Larson 1988 *The Sociology of Social Problems*. 9th ed. Englewood Cliffs, NJ: Prentice-Hall.

Israel, Joachim 1971 *Alienation: From Marx to Modern Sociology*. Boston: Allyn and Bacon.

Jones, Ron 1981 *No Substitute for Madness*. Covelo, CA: Island Press.

Kerbo, Harold R. 1983 *Social Stratification and Inequality: Class Conflict in the United States*. New York: McGraw-Hill.

Kozol, Jonathan 1988 *Rachel and Her Children: Homeless Families in America*. New York: Crown Publishers.

Kushner, Harold S. 1981 *When Bad Things Happen to Good People*. New York: Avon.

Liebow, Elliot 1967 *Tally's Corner*. Boston: Little, Brown.

Lynd, Robert, and Helen Lynd 1929 *Middletown*. New York: Harcourt Brace Jovanovich.

Lynd, Robert, and Helen Lynd 1937 *Middletown in Transition*. New York: Harcourt Brace Jovanovich.

Malcolm X and Alex Haley 1965 *The Autobiography of Malcolm X*. New York: Grove Press.

Marshall, Thurgood 1979 Address delivered November 18, 1978. *The Barrister*, January 15.

Marx, Karl 1844 *Economic and Political Manuscripts of 1844*. 1964 ed. New York: International Publishers.

Marx, Karl, and Friedrich Engels 1848 *The Communist Manifesto*. 1955 ed. New York: Appleton-Century-Crofts.

Matras, Judah 1984 *Social Inequality, Stratification, and Mobility*. 2nd ed. Englewood Cliffs, NJ: Prentice-Hall.

Merton, Robert 1938 "Social Structure and Anomie." *American Sociological Review*, 3:672–682.

Mills, C. Wright 1959 *The Sociological Imagination*. New York: Oxford University Press.

Nettler, Gwynn 1983 *Explaining Crime*. 3rd ed. New York: McGraw-Hill.

Phillips, Kevin 1990 *The Politics of Rich and Poor*. New York: Random House.

Quinney, Richard 1980 *Class, State, and Crime*. 2nd ed. New York: Longman.

Ritzer, George, and David Walczak 1986 *Working: Conflict and Change*. 3rd ed. Englewood Cliffs, NJ: Prentice-Hall.

Ryan, William 1976 *Blaming the Victim*. Rev. ed. New York: Vintage.

Schacht, Richard 1970 *Alienation*. New York: Doubleday (Anchor).

Seeman, Melvin 1975 "Alienation Studies." In *Annual Review of Sociology*. Ed. Alex Inkeles, James Coleman, and Neil J. Smelser. Palo Alto, CA: Annual Reviews.

Sennett, Richard, and Jonathan Cobb 1972 *The Hidden Injuries of Class*. New York: Random House.

Silberman, Charles E. 1980 *Criminal Violence, Criminal Justice*. New York: Vintage.

Simmel, Georg 1902–1903 "Metropolis and Mental Life." In *The Sociology of Georg Simmel*. 1950 ed. Ed. Kurt Wolff. New York: Free Press.

Simpson, George E., and J. Milton Yinger 1985 *Racial and Cultural Minorities: An Analysis of Prejudice and Discrimination*. 5th ed. New York: Harper and Row.

Skolnick, Jerome H., and Elliott Curie 1985 *Crisis in American Institutions*. 6th ed. Boston: Little, Brown.

Steinberg, Stephen 1989 *The Ethnic Myth: Race, Ethnicity, and Class in America*. 2nd ed. Boston: Beacon Press.

Straus, Murray A., Richard J. Gelles, and Suzanne K. Steinmetz 1988 *Behind Closed Doors: Violence in the American Family.* New York: Doubleday.

Sykes, Gresham M., and Sheldon L. Messinger 1960 "The Inmate Social System." *Theoretical Studies in Social Organization of the Prison,* Pamphlet 15. New York: Social Science Research Council.

Terkel, Studs 1972 *Working.* 1975 ed. New York: Avon Books.

Thrasher, Frederic 1927 *The Gang.* Chicago: University of Chicago Press.

Turner, Jonathan H., Royce Singleton, and David Musick 1984 *Oppression: A Socio-History of Black-White Relations in America.* Chicago: Nelson-Hall.

Van den Berghe, Pierre 1978 *Race and Racism: A Comparative Perspective.* New York: John Wiley.

Wallerstein, Immanuel 1974 *The Modern World-System.* New York: Academic Press.

Weber, Max 1904–1905 *The Protestant Ethic and the Spirit of Capitalism.* 1958 ed. Trans. Talcott Parsons. New York: Scribner's.

Weitzman, Lenore J. 1985 *The Divorce Revolution: The Unexpected Social and Economic Consequences for Women and Children in America.* New York: Free Press.

Wilson, William 1987 *The Truly Disadvantaged: The Inner City, the Underclass, and Public Policy.* Chicago: University of Chicago Press.

Yankelovich, Daniel 1982 *New Rules: Searching for Self-Fulfillment in a World Turned Upside Down.* New York: Bantam Books.

Zimbardo, Philip 1972 "Pathology of Imprisonment." *Society,* 9:4–8.

Does the Individual Really Make a Difference?

An Introduction to Social Change

*W*hen I was an undergraduate at the University of Minnesota, I did not understand much about government and politics. I knew I lived in a democracy, but I did not seek to understand exactly what that meant. I was suspicious of communism, but I did not understand exactly what it was. I looked forward to the day I would be 21 so I could vote, although I did not know how I might do so intelligently. Looking back, I was truly a naive undergraduate, but at least I really wanted to understand politics.

In the library one day I met another student who looked old enough to vote, so I asked him if he had voted in the last election. His answer was a simple no. I was shocked, and so I began to question him. He presented his case to me: "It doesn't matter if I vote or not. One vote will never make a difference to anything." I pointed out that if everyone thought that way, our democracy would be a farce. He replied that he was not talking about "everyone"; he was talking about himself. I asked him about his influence on others around him: family, friends, other students. He replied that they did not know whether he voted; they could vote if they wanted.

So began my investigation of democracy. Through all my reading and discussions, the memory of this man from the basement of Walter Library keeps haunting me. I have broadened it considerably: What difference can one individual make on others around him or her? What difference can the individual make in any established group, formal organization, community, or society?

Much of the sociological perspective focuses on the power of society, social forces, social patterns, a world out there that influences, shapes, and controls the individual actor. But how does this

201

world change? Surely each of us has something to say about those forces. Does my vote matter in democratic society? Do our votes collectively matter? Surely some individuals shape society—don't they?

What is the real relationship between the individual and society? Of course we are shaped by society; but do we, in turn, shape society? Do individuals really make a difference? It is relatively easy to establish that most individuals have some influence on those with whom they interact. Some of us make a great difference. It is more difficult to see how individuals can influence established organizations. It is far more difficult to see how they might influence society.

It is fashionable in the United States, of course, to believe that individuals can do anything they set their minds to: if someone wants to make a difference to other people, to society, even to the world, he or she can. It is also fashionable to believe that society changes because of the efforts of the individual. To believe these things does not make them true; to believe them is really a statement of faith taught in society. It is very important for people to believe that they make a difference: "I'm important." "My life matters." "I do have an effect on the lives of others." "I can shape the future of society."

This chapter will examine the power of the individual to change society. We will start small, however, and look first at how individuals can make a difference in:

- ☐ their own lives
- ☐ the lives of those with whom they interact
- ☐ various organizations to which they belong

Then, finally, we will examine society. As you will see, each topic is highly complex, and you will probably find that the actual difference the individual can make is less than you imagine.

The Individual's Influence on His or Her Own Life

If we are free at all and exercise control over our own ideas, values, actions, and directions, each actor makes a difference *to his or her own life*. I influence what I do. I make decisions that allow me to go one way rather than another, to believe one thing rather than another, to act a certain way, or to become a certain type of person. We examined the question of freedom in Chapter 6, especially the claim that we do have some control, some ability to make a difference in directing what we do with our own lives. In the context of this chapter, we can say that if we are free, our lives do matter—they matter to ourselves. We make a difference in our own lives.

But because of the many social and other factors that control us, this ability is always limited, often to a great extent. We always act in a social context. Social patterns always matter. Role, class, culture, and institutions always guide our decisions and our lives. Our problems are always linked to social problems, and our successes are linked to the state of society. The sociological view leaves some room for individual freedom, but not a great deal.

When people ask if the individual really does make a difference, however, they are usually asking about *impact on others*. Do our acts matter in terms of influencing other individuals, groups, communities, or societies? The topic really becomes one of *social change*: Can my acts influence others—do they matter to anyone besides myself? In the wider social context, can the individual really influence the direction of society in a way that he or she chooses? Or is social change really the result of other factors that have nothing to do with individual actions, that have mostly to do with external forces outside individual control?

The Individual's Influence on Other Individuals

We can begin to understand the importance of the individual in society by first looking at social interaction. Do individuals make a difference in how other people live their lives? Do individuals influence their children and friends? Are they able to influence the lives of other individuals trapped by poverty or ignorance?

The Problem of Measuring Influence

Let us start rather simply. In all that we do, we encounter others who follow different interests from ours, who have different views, different priorities, different problems. If we try, can we make a difference to them? The question of the influence of one actor over another is difficult to answer. Teachers, for example, often exaggerate their own influence over students: "I taught my class the locations of all the nations of Africa. They really know something." Or, "Now that I've taught my class the harmful effects of drugs, they know that if they do drugs, they'll destroy themselves." But such influence is much less than imagined.

First of all, people often forget most of what they are taught because it is not useful to them. Second, many others are influencing us at the same time, so the influence of any given individual must be understood in the context of all these other influences. The problems of my family may be far more important to me than what I am trying to learn about geography; the desire to be accepted by my friends may be far more important to me than a lecture on drugs by someone I barely know. Third, the influence of a given individual over someone else is often unintentional: the teacher may be teaching the location of nations, but the student may be learning to hate geography; the teacher may be teaching the harmful effects of drugs, but the student may be learning that adults are hypocritical, unrealistic in their expectations, and not to be trusted. Of course, the influence can be unintentionally positive: the student in geography may become excited about visiting other societies; the student learning about drugs may be influenced to confront his or her parents about their alcohol abuse.

I remember Tom Costello, a student I taught in high school, coming back to school many years after graduation and declaring to me: "Oh, you're Mr. Charon—the one who was always making corny jokes in class." He remembered me!—but for reasons I never intended. Some students come back and tell me: "You're the one who used to take your shoes off when you lectured" or "You're the one who used to be late to class because you always had to have a cigarette." I, of course, wanted to be remembered as the one "who taught me something," or the one "who loved learning," or the one who "cared about his students."

Successful Influence

Of course, we do have an effect on others. Most of the time it is fleeting, but sometimes it is lasting. Most of the time the effect is unintentional, but sometimes it is intentional. George Herbert Mead called others who are important in our lives "significant others," and almost all of us are significant others to some people. The love we give spouse and children almost always affects their lives in a positive manner. Parents, teachers, friends, employers, rock stars, members of the clergy, and political leaders are significant others whose lives become examples to us, whose ideas or values become our own, whose perspectives we use to understand our environment. Sometimes an encounter with someone will change our whole direction in life, and what may at first seem to be a small influence becomes a huge influence. A chance encounter with a book salesman altered my life considerably; without that encounter I probably never would have tried to write a book, and my life and research might have gone in another direction. The reading of this book is also but a chance encounter for its readers. Be realistic: How much lasting influence can I have? For most students, not much. For a few, maybe I can influence their direction. For a few, maybe I can teach something useful that will last for a short while. For a very few, maybe I can influence them to have a love of learning or a love of sociology or a concern for the fate of humanity that will remain important all their lives. Unfortunately, I also fear that for some I will supply ammunition for being intolerant, sexist, racist, anti-intellectual, or antisociology—unintentional influences, but nonetheless real.

To believe that the individual matters and can influence others must take into account negative influence, too. Those of us who abuse our children will lay the foundations for misery—for our children and for those with whom they interact—and that misery may last their whole lives. Those of us who teach our children that it is all right not to care about others, to hate those unlike ourselves, and to harm or exploit them will also affect them. Those of us who hurt others may cause emotional harm, and that, too, will make a difference to their lives.

Influence over Other Individuals Exists Within a Larger Social Context

The influence of one actor over another actor cannot be understood in isolation. It does not take place in a vacuum but always exists in a larger *social context*. The likelihood of influence depends on larger social trends. My likelihood of influencing you to enjoy opera as I do depends on whether opera is loved in society and whether it is rapidly being replaced by rock music. It depends on whether we live in Italy or the United States; it depends on whether we live in the 1990s or the 1890s. The teacher influences students who are ready and interested. The religious leader converts individuals who are seeking conversion. The political leader influences individuals who agree with his or her political philosophy or, sometimes, who are rebelling against their parents' political philosophy. A young person is influenced by a movie star who represents the culture to which that young person belongs. The individual who influences someone to try drugs is aided by a social context that regards drug taking as acceptable behavior.

Thus, the influence of one actor over another may be real, but it is often exaggerated and is always facilitated or made more difficult by the social context. All of us make a difference sometimes. That difference may be in the direction of tolerance, love, caring, and growth; it may be in the direction of intolerance, hatred, and destructiveness. There is usually no way of knowing whom we have affected or how strongly. For most of us, interpersonal influence remains an article of faith that we believe in.

Influence Is Two-Way

Finally, it is important to recognize that interaction is two-way. I may influence others through what I do and say; they, too, may influence me. We negotiate influence; usually, no one has complete influence over the other—it goes both ways. My attempt to influence my wife will usually be met with her attempt to influence me back. The end result is that I do not get my way, and she does not get hers. We both make a difference to each other, but it is almost always less than we would like, because neither of us is simply an empty vessel. To see influence as a one-way affair exaggerates the importance of the individual in the interaction.

The Individual Versus Social Organization

To affect ourselves or to influence other individuals is one thing; to affect a group, formal organization, community, or society is something else again. Imagine the difficulties involved in accomplishing something very important and lasting in an established social organization.

My first job after graduating from college was as a high school history teacher, and I was prepared to make a difference in the world. I remember that I wanted to contribute to all humankind. Perhaps I would teach someone who would become a great leader; perhaps I would teach ideas that would spread; perhaps I would be recognized as a model teacher for all to learn from. It wasn't clear how I would do it, but I knew I wanted to make a positive mark on society. A short time after I began, it became obvious to me that my sights were set too high. At least I could have an impact on the community of St. Paul. After teaching for a month or so, I knew that if I wanted to have an impact I would have to settle for Harding High School. My idealism changed when I realized how few students actually took my classes or knew who I was. Like most idealistic teachers I eventually came to realize that my real chance for making a difference in people's lives was to be found in the everyday interaction with 150 individuals I met five hours a week. Yet I came to see that I could not expect great things there either. I was not an important person to many of these people. Some saw me as an intruder in their lives, and some did not often understand what I was trying to teach. My actions in the classroom did, in fact, eventually influence several students, but I am afraid I had a really lasting effect on only a few and, even then, often not in the direction I had intended. I never did influence society or St. Paul, and I left Harding High School as I had found it, having had a minimal impact on it.

I really wanted to have an impact on an organization. Why couldn't I? What stood in the way? Why is it so difficult—perhaps impossible—for the individual to have a real impact on an organization?

The Individual Confronts Social Patterns

We return to the existence of *social patterns*. Every organization eventually develops certain ways of doing things. That's actually what is meant by being organized. People know what others are going to do, and they understand what they are supposed to do. Structure distributes positions in an organization, which are usually ranked and have attached to them roles, or scripts, laying out what is expected. Culture is taught to all, creating a shared set of beliefs, values, and rules that guide actors as they interact. In society there are institutions, long-established procedures that guide the individual. For example, people in American society have established marriage as an institution that the vast majority of us follow. When we marry, the general outline for what we are to do is laid out in advance: courtship; engagement; a religious ceremony; a reception with friends and family; an agreement to be loyal and to give love, money, and time to one another; and so on. In our individual marriages we establish our own social patterns over time: who does what, the degree of independence each of us has, how much money we should spend versus how much we should save. As new problems arise in our marriage, we have to discover new solutions, and they, too, become established as patterns. As we have children, change jobs, and move from one community to another, we have to alter what we do, but we revise the old patterns rather than simply throw them out. The revisions become new patterns we follow. The bond between us depends on our feelings, but it also depends on all of the patterns we have come to share. For one of us to decide to abandon these patterns or to change them radically upsets the relationship. And if one tries to change things without consent, the danger of dissolution grows.

If it is so difficult to change social patterns in a marriage, consider how much more difficult it is in a larger social organization: a group of friends, a school, or a society. A lasting group—a class of thirty students, for example—establishes patterns early, often ones that have been developed elsewhere to guide all such classes or have been established through the demands of the teacher or even through day-to-day interaction. Any individual, including the teacher, who tries to change those patterns radically threatens the organization of the class and its success. Even the students are aware of this, and once the patterns are established, they, too, work against the rules being changed, either by the teacher or by a newcomer in the class who wants them to fit his or her own needs.

A football team works the same way. Game rules and league rules govern team play. Individuals are discouraged from openly challenging them. The team itself develops special plays, procedures for play calling and substitutions, and even subtle ways of shifting what individuals do as a play unfolds. Individuals who decide to go their own way threaten the team's success.

On the opposite end of the organizational spectrum stands society. Society has a long history. It precedes every living actor, and it will be there when every living actor dies. Its patterns—social structure, culture, and institutions—have been established over many years, and these patterns confront the individual as a generally accepted reality. The individual can cry out, "I won't do what you want of me!" and can leave that society, no longer to be influenced by it. Quiet nonconformity is possible. But it is something else for the individual to try to change the dominant social patterns. To change society is to threaten the continuation of the world as it exists for most people.

Social patterns, according to Emile Durkheim, take on a life of their own. They exist "out there" someplace—invisible, real, external to us, influencing and even controlling us. When we break these patterns, we challenge their reality. See what happens when we decide to go it alone in social organization, to refuse to follow the procedures, rules, truths, and values that were established long ago. We are talking about not only laws but also many other patterns that guide us in every move we make, from greeting someone on the street to burying the dead. Actions are neither spontaneous nor random. They generally follow patterns laid out by strangers long dead.

What chance, then, does the individual actor have against social patterns? We each live in a social reality that others have become used to and generally are fearful of losing. No matter how much we might dislike our situation, there is something in most of us that cries out to keep the structure, culture, and institutions that we have. We might hate society as it is, but it is the only world we know.

The United States is a society with a system of government developed for more than two hundred years. It is something defended by almost all who are appointed or elected to positions within that structure. It is laid out in a constitution that most of us respect. We are used to the way it works. And, for all of its faults, it does work, at least for large numbers of us. And if we do not like it, who among us can change it in any dramatic way? If we work within the government, it may be possible, but we may threaten our own rise in that structure. It also may be possible if we work outside of government, but we may threaten our own comfort, our occupational success, and sometimes even our lives.

This is true of a university, too. Go change it! Change its patterns: how it operates, how it is structured, what is valued and believed there, or its basic rules. It can be done, but it is very, very difficult.

Individuals can make a difference to the successful operation of an organization—within the bounds of its social patterns. They can help the organization achieve its goals or can make a difference in the opposite way by blocking those goals. An outstanding quarterback can pull together a bad team. A good president can lead society in positive directions. An outstanding businessperson can turn around a struggling company. Such individuals can make a difference—sometimes a big difference. Normally that kind of success does not change the social patterns in that organization (the individual is simply an outstanding actor within the established patterns). The individual will have a much more difficult time making a lasting and dramatic impact on the social patterns themselves.

In general, most people who work for change in an organization work for minor change within the existing social patterns. Such change may be very important, but it does not constitute lasting, significant change. Some individuals will make a difference, but as outstanding actors following a written script.

There are individuals who truly shape social patterns, however, and leave a great mark. Mikhail Gorbachev is an individual who rose up in a political structure first established in 1917. He rose because he was perceived to be someone who represented and stood for the structure, culture, and institutions established in Soviet society. He also used those social patterns to rise to the top. Normally when one rises to the top, one becomes increasingly supportive of the system that made possible that rise. But Gorbachev was different—he made a big difference to the Soviet Union and to the world because, once in a high position, he turned around and criticized the political institutions he was leading. He criticized the nature of the economy that favored people in his position. He questioned the necessity for the massive military system that formed an integral part of Soviet society. Here was a man who could really make a difference: powerful, critical, willing to make basic changes in the patterns of society. Gorbachev was an unusual revolutionary, in that most revolutionaries exist outside the dominant social patterns. He came from within the system itself: He had position, powerful allies, and intelligence. He calculated well, and he brought about great reforms. He made a difference, a great difference. This was not easy to do, nor is it common. The dead hand of the past weighed heavily on him. Others who favored a system that clearly benefited them attacked him. As unproductive and unresponsive as the political and economic system was, it was all that many of the people had known, and it was natural to question if a new system would necessarily be better. Everything seemed to oppose any chance that one man would make so much difference. In fact, it is probably safe to say that his influence was even greater than he imagined it would be. As a result of his efforts, change took on a life of its own, and many events he did not originally intend occurred. Eventually he fell from grace, to be followed by other leaders, each trying desperately to deal with tremendously difficult problems, each trying to establish new institutions, each finding it difficult—even impossible—to change inherited institutions. History, of course, is complex, and the case of Gorbachev has not yet been fully written. Despite his personal fate, however, Soviet society and Eastern Europe can never fully go back to their failed institutions, and Gorbachev will go down as an individual who made an important difference in the world. Without him the world would have changed; but the way it changed, and the speed with which it changed, are a result of his efforts.

Max Weber recognized that some individuals could make a significant impact on society's patterns but that such individuals usually rose up outside the legitimate order. They were revolutionary heroes who gathered around them large numbers of the dissatisfied. Someone such as Gorbachev—one who rose up in the system to the very top—is an unlikely candidate for revolutionary change. Whatever people's motivation is for rising to the top (even if it is very noble), once they are there, their agendas change. When people get to high positions, they confront history and social patterns directly, and they become what they never intended. It is difficult to say no to a social structure and set of institutions that created your success. Even revolutionaries who overthrow the scoundrels usually become scoundrels themselves without really changing very much. We might vote out the other party only to find that the new party is not capable of making the real changes promised—not because its members are liars but because revolutionary change is made so difficult within the social patterns already in existence. This is the real lesson of the Clinton presidency. How much can one individual change social patterns that are deeply embedded, no matter what his or her intentions are? Political candidates promise change, and many really believe they can bring it about, only to meet the realistic power of our inheritance as a society.

The Role of Social Power

What most of us fail to consider when we think about the influence of an individual on an organization is the element of *social power*. The individual can change an organization only if he or she has social power. Power means the ability to achieve one's will in relation to others. One's ability often arises from *high position* in an organization, but it also arises from attractiveness, large numbers of followers, wealth, weapons, intelligence, or persuasive ability. Parents have great power over children, and thus they can influence their directions, their ideas, their values. As other elements compete with that power, parental influence lessens. Corporate leaders have great power, and they use it to shape policies in their interests. Sometimes it is to change society (the tax structure, the relative power of unions, the degree of governmental "interference"); more often it is to protect the social patterns as they operate (private enterprise, inheritance practices, a court system based on the ability to pay). The president of the United States has great power and thus is able to have more influence on the direction of society than the rest of us. (After all, what do we have that allows us to influence that direction? A vote? A contribution to one official? Our going door to door trying to get votes for our candidate?) A skillful leader in a well-armed, well-funded revolutionary group may have great influence on society, and a skillful religious leader may have an impact on a congregation of believers or even on the direction of society. In understanding how such influence is possible, however, it is important to recognize not only that desire is necessary but also that desire must be wedded to *power*. Ideas can be effectively challenged only with power. Criticism must be backed up with power. New directions for an organization must result from those who have more than good ideas or good intentions, and new social patterns can arise only from people who have the power to institute them.

Power is a complex matter. First of all, those with the most power are usually those who do not want basic change in society. They benefit the most from social patterns as they are. They rose up through those patterns, and they will generally approve of those patterns. Second, one actor's social power is one part of a *social equation*. Power is exerted from both sides. Even if a person desires change and has power to bring it about, those on the side protecting the social patterns will also have power, and in the vast majority of cases their power will be much greater. Part of every organization (from the family to the society) has mechanisms for dealing with those who rise up to try to change that organization as it exists. To be able to change society is to have enough power to influence those who defend society as it exists.

Change usually occurs, therefore, not because of the efforts of one individual but because people work together, form a power base, and bring about change. A leader cannot change society alone; by definition, he or she needs a power base that includes other individuals willing to work in the same direction for change. In 1789, the workers and middle-class people of France united around emerging leaders and overthrew the monarchy. In the 1950s, African Americans in the South organized around Martin Luther King and together began to bring the system of segregation down. King was important as a leader, but alone he was without power in society. *Social movements*—loose organizations of large numbers who can be effectively mobilized around leaders to march, protest, boycott, strike, and actively confront the opposition—change society. More organized protest groups—Greenpeace, Mothers Against Drunk Drivers, Amnesty International—change society. Various individuals within them may make a difference in the direction of change, but it is many people working together in opposing established social patterns who have the real potential for changing society. Each individual is a resource used to bring about change. Alone they are not influential, but united they can make a difference.

Success is never guaranteed, no matter how much one desires to have an impact, no matter how much power one has, no matter how noble one's cause. As a result of all our efforts, four possibilities exist: (1) the social patterns still may not change; (2) the social patterns may change because of the organized opposition, but in a direction unintended (perhaps toward more oppression); (3) the social patterns may actually change in the direction desired; or (4) the social patterns may change in exactly the way desired. Number 1 is the most likely to occur; number 4 is the least likely.

Social Change: A Sociological View

If sociologists recognize that the individual does, in fact, make a difference sometimes in changing a social organization, but that this difference is usually minor and often unintentional, how else do they approach the problem of social change? If the individual is not primarily responsible for change, what is? It is probably best to begin answering this question by listing and explaining five guiding principles that most sociologists believe.

Change Is Inherent in All Social Organization

The first principle is that every organization naturally changes. As its size changes, it changes; as it becomes older, it changes; as its environment changes, it must adjust. Each event in its history becomes part of its past and can be recalled and used to make decisions later on. Indeed, it is erroneous to see social organization as a rigid, permanent set of social patterns that represents enduring stability and order. Every action of every individual alters society a tiny bit. Every decision by our government alters society a tiny bit. Of course, some actions, individuals, and decisions are more important than others. But the point is that society never stays the same from moment to moment. Change is to be expected. Whatever one likes about society today—its music, its movies, its family patterns, its level of religious commitment—will inevitably change. Whatever exists today will be at least slightly different tomorrow.

Societies even have rates of change. Some societies change far more rapidly than others. The rate of change is an integral part of society; it is one of the social patterns that exists. Events or individual acts may slow down the rate for a short while or speed it up, but over the long run the rate is predictable. The rate of change in the United States in the twentieth century has been extremely high compared with those in most other societies in such areas as increasing urbanization, the rise of a service economy, egalitarianism within the family, and geographical mobility. These rates have dramatically accelerated since World War II. The United States is a very different society today than it was in 1900. Some societies—Japan, for example—have had even higher rates of social change than the United States. Others, such as Iran, India, and China, have had lower rates.

Adults continuously remind the young that "the world is a lot different today than it was when I grew up." Well, look around you right now. The society you see will be very different ten years from now. Whatever you like may be gone. Whatever you do not like may be improved, or it may be worse. Everything changes, but the change is generally gradual, and it is usually impossible to pin down one individual or group as responsible for the change.

Change Stems More from Social Conflict than from Individual Acts

The second sociological principle is that change is probably more a result of social conflict than a result of the acts of any individual. Organization is never as peaceful and finished as it appears on the surface. There is always disagreement, always protest. It is when the authorities say *no* and when others continue to say *yes* that conflict and social change arise. Rarely do such individuals have their way, but because they fight for what they believe or want, some change occurs over time, often in ways they did not even intend. The civil rights movement in American society has never achieved its goal of racial equality. However, because of the conflict that it generated— the back-and-forth struggle between the movement and those who supported the institution of segregation, the culture of racism, and the unequal racial social structure—patterns changed gradually in the direction that the movement cried out for, even though never to the point that it desired. The conflict has given rise to greater educational opportunity, more equal political and civil rights, and more equal economic opportunity at the top of society. However, it has not made much impact on the anger and hopelessness among those African Americans who have no real economic or educational opportunities; on the increasing numbers of young, single minority parents; and on serious inner-city drug abuse. Did the civil rights movement succeed? Yes (partially): society has become more open to middle-class African Americans, and the political and legal system is more sensitive to problems of racism. And no: equality does not prevail, racism is still widespread, and for large numbers, nothing has really changed for the better.

Karl Marx emphasizes the role of social conflict as the source for change. History is the struggle of opposing classes, he writes. Society is made of workers and owners, and over time the inevitable conflict between them alters society. For a long time the open conflict is kept in check, and then suddenly a great upheaval brings down the old and creates the new. The new society, according to Marx, is a synthesis: it is the coming together of the old and the new, those who fight to keep what they have and those who are opposed and must fight for their rights. The new arises out of social conflict. Marx sees the English and French revolutions as examples of mass conflict that created such syntheses: they were examples of societies moving from feudalism to capitalism. The revolutions were really culminations of social conflict that existed for hundreds of years; the new societies were significantly different from the old, but not brand new, because they were created out of social and economic trends existing in the old.

Most sociologists see conflict as inevitable. As long as there are people with different ideas and different interests, there will be conflict; and as long as there is conflict, nothing will stay the same. Everything is in flux; nothing is inevitable, and everything is open to challenge and change.

Like Marx, Max Weber (1924) sees social change as arising out of conflict—between those who defend the traditional order and those who act against that order. Those who rise up are revolutionary, they gather followers, and sometimes through their efforts the old order is overthrown. But it is never completely overthrown, because some old patterns help forge the new ones. New conflict follows immediately between those supporting the new patterns and those who oppose them. The new eventually loses its newness and becomes *tradition*, and new charismatic leaders eventually attract followers and fight once again. History is the struggle between tradition and revolution.

Rather than focusing on the individual or the protest group in isolation, sociologists tend to focus on what happens *among* people as a source of change. Individuals matter; but more than that it is individuals and groups in interaction with one another that matters. And it is the conflict generated in that interaction that brings about lasting change.

Change Is Most Likely when the Social Situation Favors It

A third principle is that individuals, groups, and social conflict are most likely to change an organization when the social situation favors it. Hitler is an example of an individual who made a great difference in history. He significantly changed the social patterns in Germany to accommodate a totalitarian dictatorship, his efforts to establish German supremacy clearly led to a world war, and through his influence millions of people were killed. Fifty years after Hitler's death his influence is still felt the world over. Many individuals and groups are still attracted to his philosophy and hold him to have been a great leader. Much of the world sees him as a representative of all that is evil in human beings and all that is possible for an all-powerful individual to attain. Most of us would admit that Hitler truly made a difference: part of it was intentional, part unintentional.

Hitler, however, was not successful simply because he wanted to change the world. He was a part of history as much as a leader of history. He was a product of German society as much as a molder of German society. Without the right social circumstances, he would not have had the impact he did.

Scholars of history remind us of some of the most important reasons that Hitler was able to come to power: the humiliating peace treaty that ended World War I; the Great Depression, which

devastated the German economy; and Germany's paralyzed government, plagued by extremism from all sides. Hitler was a product of
many social patterns in German culture, and he appealed to these in
his rise to power: German nationalism, militarism, authoritarianism,
and anti-Semitism. In large part he came to power and made a difference because he tapped into and used German cultural patterns.

Hitler also made a great impact on the world because Germany
was highly bureaucratized and scientifically advanced. He was able
to use bureaucratic principles to organize society, control the population, build an efficient military machine, and transport, imprison,
and systematically murder millions of people. He was able to use the
German scientific community to develop weapons of war that were
often superior to those of his enemies.

Without this social context, Hitler would not have risen to
power, and his influence on German society and the world would
have been impossible. So it is with every influential leader in history.
The individual makes a difference when social conditions are right:
Luther, Lenin, Mao, Roosevelt, Lincoln, King, and Gorbachev are individuals who made a difference because society was ready for them.

Ideas sometimes change society, but ideas, too, exist within a
social context. To create new ideas the individual must build on what
is known. Great ideas are often a synthesis of the ideas of others or
the reactions to these other ideas. Newton, Galileo, Copernicus, Dar-
win, Freud, and Marx are all thinkers who revolutionized how people in society thought, but their ideas were built on those that had
gone before. Moreover, the influence of these ideas depended on
conditions in society that encouraged their promulgation. It is not
only the "truth" that wins out in society (it may or may not), but
those ideas that have sponsors: groups, communities, social movements, or classes willing to believe and to sell them to others. The
ideas of some individuals make a difference, but there is always a social context that helps determine their acceptance or rejection.

Weber's emphasis on the importance of charismatic authority
in history underlines this point. Those who make revolution tend to
exist within certain periods of history when the old world is collaps-
ing, when institutions no longer work well, and when old ideas no
longer seem sensible. They rise up because others look to them in a

world where many are dissatisfied. In short, revolutionary individuals make a difference only in a much larger social context that is ripe for their influence. It is probably true that revolutionaries always exist; they come to make a real difference only when the times are such that others are ready for them. They make little difference when few are willing to listen.

It is tempting for the sociologist to discount the impact of the individual on society. Certain individuals do influence other individuals and do change social patterns. Their impact is sometimes great and cannot be discounted. Yet it is vital to always see these individuals in a larger social context that (1) helped produce them and (2) made their influence significant. This same point can be made of any social organization, be it a group, formal organization, or community: certain individuals can make a difference, but they make a difference within a wider social context.

Most Change Results from Social Trends

A fourth principle that most sociologists subscribe to is that much of what we call social change is the result of impersonal social trends over which individual actors have little control. A *social trend* is change that arises from the actions of many individuals who are attempting to deal with their everyday situations and who act in a similar direction and produce a cumulative effect on society. Few people actually intend to change society, but together their acts do, in fact, cause a change. So, for example, many people today are putting off marriage until they get older; many are deciding to divorce; many are remarrying after getting divorced. These are social trends, very general tendencies shared by many in society. As a result, society changes. Social trends are themselves caused by even larger trends such as industrialization and increasing individualism. They create change in spite of the fact that each actor's influence is unintentional. Important general trends in our society today might include population trends (fertility, mortality, and migration), urbanization, industrialization, increased use of technology, and bureaucratization.

The United States, for example, is experiencing a revolution in computer technology. Indeed, the development of technology—the application of knowledge to solving human problems—has been a rapidly accelerating social trend for at least three or four hundred years. Computers are altering every aspect of our lives, from education to music, from diagnosing illness to making war.

Social trends are long-lasting, far-reaching, general developments that affect all the various social patterns in society. In the long run, such trends are the most important forces for social change. They set an almost irreversible direction for society. Individuals normally contribute to social change if their acts and ideas are consistent with these trends. Many in society may hate such trends and fight them, but these trends have an inertia; once begun, they take on a life of their own and are difficult to turn around.

Weber (1904–1905) maintains that a social trend he calls "the rationalization of life" is dominating Western societies. Throughout society, he writes, there is increasing reliance on "calculation," "efficiency," "problem-solving," and "goal-directed behavior." This is the meaning of modern life to Weber: instead of tradition, human beings value reaching *goals*—organizing themselves most efficiently, making and selling goods in the most profitable way, and calculating the most effective way of getting what they want. "That's the way we have always done it" is replaced by the ethic of "This is the smart way of doing it." Tradition is not valued; achieving our goals is. Indeed, Weber argues, we are no longer a people committed to value-oriented behavior, either: I do what I do not because of commitments to values (such as knowledge, goodness, equality, love, and freedom), but because my behavior is the most rational way of achieving my ends. Weber documents the declining importance of tradition, values, and feeling as human behavior becomes increasingly rational and calculating. Society is becoming efficient in many diverse ways: we are able to turn out millions of television sets, tons of wheat, and large numbers of college graduates. We are able to provide health care to more people than ever in the history of mankind, and we are able to encourage more people to buy more goods without any cash. We can provide more answers than ever before as science and mathematics dominate our society, and we find our lives less and less private as computers and bureaucracies are able to monitor what we do.

This is the most important modern trend, according to Weber. Many sociologists agree. Once begun, it is difficult for any individual to turn it around. Individuals may matter in society, but compared with this trend, their influence is minimal. Like most other trends, the rationalization of life is a mixed blessing: it contributes to a better life, but it takes away something important. The rise of calculation challenges mystery and myth. The rise of bureaucracy threatens the small entrepreneur. The dominance of mind deemphasizes feeling. The desire to find the best way of doing something erases the past.

Other trends are identifiable, too, each one important, each one the work of many thousands of individuals going about their business in life, trying to rear children, make enough to live on, and do what they have to do. No one person has much impact alone, but together they contribute to the trends. Does the individual really make a difference? It is difficult to say yes when we look at these general trends.

Yet even against remarkable odds, individuals sometimes rise up and make a difference through challenging the trends and establishing their own. The best example in our lifetime is probably Ayatollah Ruhollah Khomeini in Iran. He led a social movement that overthrew the shah, who had stood for the rationalization of life (including the liberation of women, secular education, industrialization, and technological development). The revolution was fought in the name of tradition, Iranian independence from the West, and a return to religious values. At least in the short run it has succeeded. It reversed what seems to be an almost irreversible social trend. Yet it may be simply a rear-guard movement. Over the long run the rationalization of life in Iran may resume.

Social Patterns Persist

Our final principle is that there is a strong tendency for social patterns to hold on. Think for a moment what a social pattern is. People interact, and over time they develop routines (for example, rules, expectations, and shared views). Those routines come to be established and become an integral part of social interaction. One such pattern is that some people become more powerful than others;

some have more privileges and prestige than others. A related pattern is that roles are established, expectations about how people are supposed to act in their various positions. Some patterns are best thought of as rules—procedures, informal expectations, morals, customs, and laws—and these become taken for granted in the interaction. People share basic value commitments and a set of beliefs about the nature of reality. They establish certain ways of handling problems. Such patterns characterize all social organization. The longer and more intense the interaction, the more important and established become such patterns.

Such patterns tend to hang on. This is the way we have always done something. This is the way we have always thought. These are the rules we have always believed. The past acts as a force for right. Furthermore, those in society who are relatively well off will spend money, life, and time defending such patterns, which they honestly believe are right. In fact, most of us, no matter how critical we are of the social patterns that make up our lives, fear change, because it may threaten the existence of social organization itself. We hold on partly because we fear we will lose everything if we challenge these basic patterns.

The individual can have an impact on other individuals. The individual can have an impact on the direction of an organization if he or she works within the patterns of that organization. But basic change—change in the social patterns of organization—is very difficult to achieve, and it generally occurs for reasons other than the intentional acts of an individual.

Some Implications for Living

The last words Weber spoke were, "The truth is the truth." It was not that he died thinking he knew the truth; far from it. He, more than most others, realized that truth was extremely difficult to know. Instead, his statement reflects his commitment to seeking truth rather than security in ignorance. He understood the discomfort of many ideas.

A society develops ideas over a long period, and they become imbedded. These ideas we sometimes call culture. It is a people's way of thinking about reality. To grow up in the United States is to confront a set of "truths" taught through our various institutions. These ideas may not be true, but they are still important to us. One of these ideas is that "individuals can do whatever they set their minds to." This notion is obviously false, but it is an important article of faith for many people. We believe not only that individuals can accomplish what they choose but also that they can have a great impact on others if they choose. Our view of social change tends to be simplistic because of our culture. It is important in our society to focus on the individual rather than on something abstract like social forces; it is important to believe that the individual is responsible for change, rather than social trends that no one individual can control.

It is comforting for me to believe that I matter, that what I do will affect others in ways that I want. From the point of view of social justice, it is important for me to believe that my acts will make a difference. That is how the civil rights movement was able to achieve what it did: people had a faith that each individual mattered. I like what that movement accomplished, and sitting here at my computer declaring that the civil rights movement was really the work of organized groups acting in the right social context seems cold and almost ruthless.

But that is what sociology teaches me. The truth is the truth, and I should try to understand it as objectively and carefully as possible, even though it may not be comforting.

Does sociology necessarily lead to apathy? Is one left with no hope for much impact? Does one have to go from a life of wanting to make a difference to a life of hopeless acceptance? Not at all.

Sociology leads one to take a more *realistic* look at social change and the impact that one can have on others. It helps explain who can have an impact and under what circumstances; it helps explain why such an impact is so difficult; it warns us that impacts may not be intended. It tells me that I may not be able to change society's system of inequality but that in my own personal relationships I can speak up against racism and injustice and realistically have an impact on those immediately around me. Sometimes this can be my only lasting contribution. Sociology tells me that my greatest impact will be in relation to those over whom I have greatest power. Thus, what I do in relation to my children may matter greatly to their future. It tells me that to have any impact on society there has to be power (which I do not have as an individual). So I must contribute my money wisely to movements that represent my concerns, and I must contribute my time and efforts to social organizations that are in a position to influence policy in the direction I desire. Sociology tells me to understand that change in the direction I want does not come easily and that I must balance my anger over injustice with realistic expectations. It tells me not to be fooled: real change is in society's patterns. Simply to vote out one individual and bring in another does not mean change. Simply to pass a law for or against something does not change the way in which society operates. Finally, it warns me that change is not usually in the interests of those who succeed and that if I want it, I must fight those who benefit from the social patterns that exist. In fact, I must realize that if I want change *and* I am benefiting from the patterns that exist, I will have to make some hard choices.

Far from bringing me to my knees, sociology teaches me a realistic view of the relationships between the individual, social patterns, and social change. That view gives me more confidence in what is possible through my efforts.

Summary and Conclusion

Social change is a difficult topic. Frankly, sociologists usually have an easier time describing order.

The individual actor exists within social forces, from those in intimate relationships to those in society as a whole. It is easiest to recognize that the individual may influence those other individuals with whom he or she interacts. It is most difficult to understand how it is possible for any individual to have an impact on the society and its social patterns. Some individuals, however, undoubtedly have great influence if they act within a social context that favors such influence and if they have a strong power base. It is important to recognize that attempts to influence society are countered by the power of long-standing social patterns that are normally defended by people who have a stake in those patterns.

When sociologists examine social change, they normally go beyond the influence of the individual. Change occurs in every social organization, and it is ongoing and inevitable. It arises out of organized groups and social conflict, and it tends to be characterized by general social trends that no one really controls.

The attempts by sociologists to describe the individual's role in a changing society may not be comforting to many people, but they are realistic and useful for understanding ourselves as social beings.

References

These works focus on social change and on the difficulties that stand in the way of the human being in making significant changes in society. Some of the works examine the massive society that confronts the individual.

Bell, Daniel 1973 *The Coming of Post-Industrial Society: A Venture in Social Forecasting.* New York: Basic Books.

Bensman, David, and Roberta Lynch 1987 *Rusted Dreams: Hard Times in a Steel Community.* New York: McGraw-Hill.

Berger, Peter, Brigitte Berger, and Hansfried Kellner 1974 *The Homeless Mind: Modernization and Consciousness.* New York: Vintage Books.

Blau, Peter M., and Marshall W. Meyer 1987 *Bureaucracy in Modern Society.* 3rd ed. New York: Random House.

Blumberg, Paul 1981 *Inequality in an Age of Decline.* New York: Oxford University Press.

Blumer, Herbert 1969 *Symbolic Interactionism: Perspective and Method.* Englewood Cliffs, NJ: Prentice-Hall.

Carmichael, Stokely, and Charles V. Hamilton 1967 *Black Power.* New York: Random House.

Chirot, Daniel 1986 *Social Change in the Modern Era.* New York: Harcourt Brace Jovanovich.

Dahrendorf, Ralf 1958 "Toward a Theory of Social Conflict." *Journal of Conflict Resolution,* 2:170–183.

Dahrendorf, Ralf 1959 *Class and Class Conflict in Industrial Society.* Stanford, CA: Stanford University Press.

Durkheim, Emile 1893 *The Division of Labor in Society*. 1964 ed. Trans. George Simpson. New York: Free Press.

Eliade, Mircea 1954 *Cosmos and History*. New York: Harper and Row.

Ewen, Stuart 1976 *Captains of Consciousness*. New York: McGraw-Hill.

Flacks, Richard 1971 *Youth and Social Change*. Chicago: Markham Publishing Company.

Gamson, William A. 1968 *Power and Discontent*. Homewood, IL: Dorsey Press.

Garner, Robert Ash 1977 *Social Change*. Skokie, IL: Rand McNally.

Goffman, Erving 1961 *Asylums*. Chicago: Aldine-Atherton.

Gusfield, Joseph 1970 *Protest, Reform, and Revolt*. New York: John Wiley.

Jones, Landon Y. 1980 *Great Expectations: America and the Baby Boom Generation*. New York: Coward, McCann and Geoghegan.

Kornhauser, William 1959 *The Politics of Mass Society*. New York: Free Press.

Lauer, Robert H. 1982 *Perspectives on Social Change*. 3rd ed. Boston: Allyn and Bacon.

Malcolm X and Alex Haley 1965 *The Autobiography of Malcolm X*. New York: Grove Press.

Marx, Karl, and Friedrich Engels 1848 *The Communist Manifesto*. 1955 ed. New York: Appleton-Century-Crofts.

Mills, C. Wright 1956 *The Power Elite*. New York: Oxford University Press.

Mills, C. Wright 1959 *The Sociological Imagination*. New York: Oxford University Press.

Olsen, Marvin E. 1978 *The Process of Social Organization*. 2nd ed. New York: Holt, Rinehart and Winston.

Piven, Frances Fox, and Richard A. Cloward 1979 *Poor People's Movements: Why They Succeed, How They Fail*. New York: Vintage.

Shibutani, Tamotsu 1961 *Society and Personality: An Interactionist Approach to Social Psychology*. Englewood Cliffs, NJ: Prentice-Hall.

Skocpol, Theda 1979 *States and Social Revolutions*. New York: Cambridge University Press.

Skolnick, Arlene, and Jerome Skolnick 1986 *Family in Transition*. 5th ed. Boston: Little, Brown.

Slater, Philip 1976 *The Pursuit of Loneliness*. Boston: Beacon Press.

Smelser, Neil 1963 *Theory of Collective Behavior*. New York: Free Press.

Starr, Paul 1982 *The Social Transformation of American Medicine*. New York: Basic Books.

Toennies, Ferdinand 1887 *Community and Society*. 1957 ed. Trans. and ed. Charles A. Loomis. East Lansing: Michigan State University Press.

Toffler, Alvin 1980 *The Third Wave*. New York: William Morrow.

Turner, Ralph H. 1968 "The Role and the Person." *American Journal of Sociology, 84*:1–23.

Weber, Max 1904–1905 *The Protestant Ethic and the Spirit of Capitalism*. 1958 ed. Trans. Talcott Parsons. New York: Scribner's.

Weber, Max 1924 *The Theory of Social and Economic Organization*. 1964 ed. Ed. A. M. Henderson and Talcott Parsons. New York: Free Press.

Wrong, Dennis 1977 *Population and Society*. 4th ed. New York: Random House.

Zald, Mayer N., and John D. McCarthy (eds.) 1988 *The Dynamics of Social Movements*. Cambridge, MA: Winthrop.

10

Is Sociology Important?

The Necessity for a Critical
Understanding of Society

*I*n the final analysis, it may be true that ignorance is bliss. It may be true that people should be left alone with the myths they happen to pick up in interaction with one another. It may be true that a liberal arts education that does not have immediate practical value is worthless.

Sociology and a Liberal Arts Education

I do not believe any of the ideas above, but I wonder about them a lot. One can more easily make a case for mathematics, foreign languages, writing, speech, psychology, and economics on the level of practical use. "The student needs to know these if he or she is to get along in life," the argument goes. It is far more difficult to make a case for sociology on the basis of practical use—unless, of course, by practical use one means *thinking about and understanding the world*. If a college education is ultimately an attempt to encourage people to wonder, investigate, and carefully examine their lives, sociology is one of the most important disciplines.

Note its purpose. The purpose of sociology is to get students to carefully and systematically examine an aspect of life that most people only casually and occasionally think about. It is to get people to understand what culture is and to recognize that what they believe is largely a result of their culture. It is to get them to see that they are born into a society that has a long history, that they are ranked and given roles in that society, and that ultimately they are told who they are, what to think, and how to act. It is to get them to see that the institutions they follow and normally accept are not the only

229

ways in which society can function—that there are always alternatives. It is to get them to realize that those whom they regard as sick,
evil, or criminal are often simply different. It is to get them to see
that those they hate are often a product of social circumstances that
should be understood more carefully and objectively.

In short, the purpose of sociology is to get people to objectively
examine their lives and their society. This process is uncomfortable
and sometimes unpleasant. I keep asking myself, as I teach the in-
sights of sociology, "Why not just leave those students alone?" And,
quite frankly, I do not usually know how to answer this question.
We are socialized into society. Shouldn't we simply accept that which
we are socialized to believe? Isn't it better for society if people believe
myth? Isn't it better for people's happiness to let them be?

I usually come back to what many people profess to be one
primary purpose of a university education: "liberal arts." To me, the
liberal arts should be "liberating." A university education should be
liberating; it should help the individual escape the bonds of his or
her imprisonment through bringing an understanding of that prison.
We should read literature, understand art, and study biology and
sociology in order to break through what those who defend society
want us to know to a plane where we are able to see reality in a
more careful and unbiased way. In the end, sociology probably has
the greatest potential for liberation in the academic world, for at its
best it causes individuals to confront their ideas, actions, and being.
We are never the same once we bring sociology into our lives. Life
is scrutinized. Truth becomes far more tentative.

Sociology and Democracy

The Meaning of Democracy

Liberation, as you probably realize, has something to do with democracy. Although democracy is clearly an ideal that Americans claim
for themselves, it is not usually clearly defined or deeply explored.

Sociology, however, explores democracy, and it asks rarely examined questions about the possibility for democracy in this—or
any—society. To many people democracy simply means "majority
rule," and we too often superficially claim that if people go to a vot-

ing booth, democracy has been established and the majority does, in fact, rule. Democracy, however, is far more than majority rule, and majority rule is far more than the existence of voting booths.

Alexis de Tocqueville, a great French social scientist, who wrote *Democracy in America* (1840) after having traveled throughout much of the United States in 1831, persuasively argues that democracy can exist only where certain social conditions nurture it, only where the structure, culture, and institutions of society support its existence. It is difficult to achieve, and it is easy to lose.

When I try to define democracy, I usually end up listing four qualities. Of course, any definition of democracy will be controversial, but it seems to me that these are the qualities that stand out in a democratic society:

1. *Individual freedom.* A democratic society is one in which the individual is free in both thinking and action. One is in control of one's own life.

2. *Limited government.* A democratic society is one in which the political leaders are effectively limited through voting, law, and constitution.

3. *Individual rights.* A democratic society is one in which the rights of individuals are respected and protected. No matter what the majority favors, there is an underlying recognition that it may not take away rights reserved for the individual or any minority.

4. *Equality of opportunity.* A democratic society is one in which all people have an equal opportunity to live a decent life. That is, privilege is not inherited, but instead people are equal before the law, in educational opportunity, in opportunity for material success, and in whatever else is deemed to be important in society.

This definition, of course, must be tentative, and people will undoubtedly debate the relative significance of these four qualities. If these are what democracy means, however, it should be obvious that the questions and thinking in this book—basic to the sociological perspective—are relevant to both the understanding of and the

living in a democratic society. Because sociology focuses on social
organization, structure, culture, institutions, social order, social class,
social power, social conflict, socialization, and social change, it must
continually examine issues that are relevant to understanding a democratic society. And because sociology critically examines people
and their society, it encourages the kinds of thinking that are necessary for living in a democratic society. If we look once more at the
questions and thinking in this book, *the theme of democracy stands out.*
One might, in truth, argue that the study of sociology is the study of
issues important to a democratic society.

Sociology: Understanding Democratic Society

Chapter 2 dealt with the nature of the human being and the role of
socialization and culture in what we all become. To ask questions
about human nature is to simultaneously ask questions about the
possibility for democratic society, a society built on qualities that are
not often widespread in society: respect for individual differences,
compromise, and concern over inequality and lack of freedom. The
sociological approach to the human being makes no assumption of
fixed qualities, but it has a strong tendency to see human beings as
living within social conditions that are responsible for forming many
of their most important qualities. A society tends to produce certain
types of people and certain social conditions, encouraging one value
or another, one set of morals or another, one way of doing things or
another. Conformity, control of the human being, tyranny, and pursuit of purely selfish interests can be encouraged; but so, too, can
freedom, respect for people's rights, limited government, and equality. *The possibilities for and the limits to a human being who can live democratically are part of what sociology investigates through its questions
concerning culture, socialization, and human nature.*

Those who think about society must inevitably consider the
central problem of social order: How much freedom and how much
individuality can we allow and still maintain society (Chapters 3 and
6)? Those who favor greater freedom will occasionally wonder: How
can there really be meaningful freedom in any society? As long as
society exists, can there be freedom? Those who fear disorder and

the collapse of society will ask: How much does the individual owe to society? If society is going to exist, don't we have to limit freedom and individuality? Such questions are extremely difficult to answer, but they are investigated throughout the discipline of sociology, and they push the serious student to search for a delicate balance between order and freedom. Too often people are willing to sell out freedom in the name of order; too often people claim so much freedom that they do not seem to care about the continuation of society. The sociologist studies these problems and causes the student to reflect again and again on this dilemma inherent in all societies, especially those that claim to be part of the democratic tradition. There can be no freedom without society, Emile Durkheim reminds us, for a basic agreement over rules must precede the exercise of freedom. *But the problem is: How many rules? How much freedom? There is no more basic question for those who favor democracy, and there is no question more central to the discipline of sociology.*

The question of social order also leads us to the questions of what constitutes a nation and what constitutes a society (Chapter 3). These issues may not seem at first to have much relevance to democracy, but they surely do. It is easy for those who profess democracy to favor majority rule. It is much more difficult for any nation to develop institutions that respect the rights of all societies within its borders. A nation is a political state that rules over one—or more—societies. If it is democratic, the nation does not simply rule these societies but responds to their needs and rights, from true political representation to a decent standard of living. If it is democratic, the question the nation faces is *not*: "How can we mold that society to be like the dominant society?" but: "How can we create an order in which many societies can exist?" If it is democratic, the nation must balance the needs of each society's push for independence with the need for maintaining social order. *The whole meaning of what it is to be a society, as well as the associated problems of order and independence, are central sociological—and democratic—concerns.*

It is the question of control by social forces over the human being that places sociology squarely within the concerns of democracy. Much of sociology questions the possibility for substantial freedom. Democracy teaches that human beings should and can think for themselves. Much of the purpose of sociology, however, is to show us that our thinking is created by our social life, that, although we may claim that our ideas are our own, they really result from our cultures, from our positions in social structure, and from powerful and wealthy people (Chapter 5). Even to claim that "we are a democracy!" can simply be part of an ideology, an exaggeration we accept because we are victims of various social forces. Our actions, too, result from a host of social forces that few of us understand or appreciate: institutions, opportunities, class, roles, social controls— to name only some—that quietly work on the individual, pushing him or her in directions not freely chosen (Chapter 6). Sociology seems to make democracy an almost impossible dream, and to some extent the more sociology one knows, the more difficult democracy seems. *Indeed, sociology tends to simply uncover more and more ways in which human beings are shaped and controlled. This, in itself, makes sociology very relevant for understanding the limits of democracy. It causes one to seriously wonder if human beings can be free in any sense.*

As I said earlier in this chapter, however, *sociology as a part of a liberal education is an attempt to liberate the individual from many of these controls.* The first step in liberation is understanding: it is really impossible to think for oneself or to act according to free choice unless one understands the various ways in which we are controlled (Chapters 5 and 6). For example, it is only when I begin to see that my ideas of what it means to be a "man" have been formed through a careful and calculated process throughout society that I can begin to act in the way I choose. Only when I begin to understand how powerful advertising has become in developing my personal tastes as well as my personal values can I begin to step back and direct my own life. And even then, an important sociological question continuously teases the thoughtful person: Can society exist if people are truly liberated? If people question everything, can there still be the unity necessary for order?

The study of social inequality—probably the central concern within all sociology—is, of course, an issue of primary importance to understanding the possibility for a democratic society (Chapter 4). It seems that it is the nature of society to be unequal. Many forces create and perpetuate inequality. Indeed, even in our groups and our formal organizations great inequalities are the rule. Why? Why does it happen? And what are its implications for democracy? If society is characterized by great inequalities of wealth and power, how can free thought and free action prevail among the population? If a society—in name, a democracy—has a small elite that dominates the decision making, then what difference does going to the polls make? If large numbers of people must expend all of their energy to barely survive because of their poverty, where is their freedom, their opportunity to influence the direction of society, their right to improve their lives? If society is characterized by racist and sexist institutions, how is democracy possible for those who are victims? *More than any other perspective, sociology makes us aware of many problems standing in the way of a democratic society, not the least of which are social, economic, and political inequality.*

This focus on social inequality will cause many individuals to look beyond the political arena to understand democracy. A democratic society requires not only limited government but also a limited military, a limited upper class, limited corporations, and limited interest groups. Limited government may bring freedom to the individual, but it also can simply create more unlimited power for economic elites in society, creating an even more ruthless tyranny over individual freedom. *Sociology, because its subject is society, broadens our concerns, investigates the individual not only in relation to political institutions but also in relation to many other sources of power that can and do limit democracy and control much of what we think and do.*

The democratic spirit cares about the welfare of all people. It respects life, values individual rights, encourages quality of life, and seeks justice for all. Sociology studies social problems. It tackles many problems, but in this book we have focused on the problems associated with human misery (Chapter 8). Many people live lives of misery, characterized by poverty, crime, bad jobs, exploitation, lack of self-worth, stress, repressive institutions, violent conflict, inadequate socialization, and alienation of various kinds. These are more than problems caused by human biology or human genes; these are more than problems caused by the free choices of individual actors. Something social has generally caused misery to occur. *Although it is impossible for sociology—or a democratic society—to rid the world of such problems, it is part of the spirit of both to understand them, to suggest and to carry out ways to deal with them.* Democracy is shallow and cold if large numbers of people continue to live lives of misery.

What does ethnocentrism (Chapter 7) have to do with democracy? Is this central concern in sociology relevant to understanding and living in a democratic society? We return to the issue of respect for minorities mentioned earlier. Ethnocentrism, although perhaps inevitable and even necessary to some extent, is a way of looking at one's own culture and others in a manner antagonistic to a basic principle of democracy: respect for human diversity and individuality. To claim that our culture is superior to others is to treat other cultures without respect, to reject them for what they are, to believe that everyone must be like us. Such ideas encourage violent conflict and war and justify discrimination, segregation, and exploitation. Sociology challenges us to be careful with ethnocentrism. We must understand what it is, what its causes are, and how it functions. An understanding of ethnocentrism will challenge us to ask: "When are my judgments of others simply cultural and when are they based on some more defensible standards (such as democratic standards)?" "When are my judgments narrow and intolerant; when are they more careful and thought out?" Even then, an understanding of ethnocentrism will not allow us to judge people who are different without seriously questioning our judgments. *Sociology and democracy are perspectives that push us to understand human differences and to be careful in condemning those differences.*

In Chapter 9 we examined social change and the power of the individual. This discussion, too, challenges many of our taken-for-granted "truths" concerning democracy. The sociologist's faith in the individual as an agent of change is not great. Democracy is truly an illusion if it means that the individual has an important say in the direction of society. But if sociology teaches us anything about change that has relevance for democracy, it is that intentionally created change is possible only through a power base. If a democracy is going to be more than a description in a book, people who desire change in society—ideally, toward more freedom, limited government, equality of opportunity, and respect for individual rights—must work together and act from a power base, recognizing that the existing political institutions are usually fixed against them. And before we go off armed with certainty, we should remember that our certainty was probably also socially produced and that through our efforts we may bring change we never intended and may even lose whatever democracy we now have. Social change is complex, dependent on social power, and difficult to bring about in a way we would like. *The sociologist will examine the possibility for intentional social change in a democratic society and will be motivated to isolate the many barriers each society establishes to real social change.*

Summary and Conclusion

Democracy exists at different levels. For some, it is a simplistic, shallow idea. For others, however, it is a complex and challenging idea to investigate and a reality worthwhile to create. If it is going to be more than a shallow idea, however, people should understand the nature of all society, the nature of power, ethnocentrism, inequality, change, and all the other concepts discussed and investigated in sociology. Whereas other disciplines may study issues relevant to understanding democracy and encourage people to think democratically, in a very basic sense this study is the very heart of sociology.

The book began with a chapter on science: researching the social world. One central point was developed: it is important to understand society without bias—that is, even about something so personal as society, human beings should try to be objective, to set aside the cultural reality they have learned, and to understand the world as it actually is. This critical evaluation of what we believe has a lot to do with freedom, because without it we are left with a cultural bias we are barely aware of and one that will influence all that we think. Democracy means that one must understand reality not through accepting authority but through careful, thoughtful investigation. It is through evidence, not bias, that one should understand. It is through open debate, not a closed belief system, that one should try to understand. *The principles of science and democracy are similar. There is no greater test of those principles than the discipline of sociology: an attempt to apply scientific principles to that for which we are all taught to feel a special reverence.*

Because it is a critical perspective that attempts to question what people have internalized from their cultures, sociology is a threat to those people who claim to know the truth. It punctures myth and asks questions that many of us would rather not hear. To see the world sociologically is to wonder about all things human. To see the world sociologically is to see events in a much larger context than the immediate situation, to think of individual events in relation to the larger present, to the past, and to the future. To see the world sociologically is to be suspicious of what those in power do (in our society and in our groups), and it is constantly to ask questions about what is and what can be.

The sociologist wonders about society and asks questions that get at the heart of many of our most sacred ideas. Perhaps this is why it seems so threatening to "those who know"; and perhaps this is why it is so exciting to those who take it seriously.

REFERENCES

Almost any sociological work might be included in the following list. However, I have tried to identify some works that focus directly on either democracy, the meaning of sociology, or both.

Arendt, Hannah 1958 *The Origins of Totalitarianism*. Cleveland: Meridian Books.

Bellah, Robert N., Richard Madsen, William M. Sullivan, Ann Swidler, and Steven M. Tipton 1985 *Habits of the Heart: Individualism and Commitment in American Life*. New York: Harper and Row.

Berger, Peter 1963 *Invitation to Sociology*. New York: Doubleday.

Berger, Peter L., and Hansfried Kellner 1981 *Sociology Reinterpreted*. New York: Doubleday.

Cuzzort, R. P., and E. W. King 1976 *Humanity and Modern Social Thought*. Hinsdale, IL: Dryden Press.

Jones, Ron 1981 *No Substitute for Madness*. Covelo, CA: Island Press.

Kennedy, Robert E., Jr. 1989 *Life Choices*. 2nd ed. New York: Holt, Rinehart and Winston.

Liebow, Elliot 1967 *Tally's Corner*. Boston: Little, Brown.

Lipset, Seymour Martin, Martin Trow, and James Coleman 1956
Union Democracy: The Inside Politics of the International Typographical Union.
New York: Free Press.

Marx, Karl, and Friedrich Engels 1848 *The Communist Manifesto.* 1955
ed. New York: Appleton-Century-Crofts.

Mills, C. Wright 1956 *The Power Elite.* New York: Oxford University
Press.

Mills, C. Wright 1959 *The Sociological Imagination.* New York: Oxford
University Press.

Moore, Barrington 1966 *Social Origins of Dictatorship and Democracy.*
Boston: Beacon Press.

Myrdal, Gunnar 1944 *An American Dilemma.* New York: Harper
and Row.

Tocqueville, Alexis de 1840 *Democracy in America.* 1969 ed. New York:
Doubleday.

Wilson, Everett K., and Hanan Selvin 1980 *Why Study Sociology? A Note
to Undergraduates.* Belmont, CA: Wadsworth.

Wolfe, Alan 1989 *Whose Keeper? Social Science and Moral Obligation.*
Berkeley: University of California Press.

11

Afterword

Should We Generalize About People?

*E*ver since the first edition of this book was published, I have become sensitive to another important question that should be dealt with. It gets to the heart of what social science is, and, in a way, it belongs in the first chapter on science. However, it also deserves its own place, so I decided to add it as an additional question to this volume. It is an important question to ponder and debate, and it took me a long time to think about it, research it, and try to answer it.

In fact, it is a topic that lies behind almost every discussion about human beings. It is implicit every time we try to understand people: all people, some people, or a given individual. It comes up whenever we try to label others or are labeled by others. It is part of every discussion of prejudice, the nature of American life, Russians, men, women, young people, liberals—in discussions that involve any attempt to categorize people. The question highlights a conflict that characterizes almost all of us: We categorize other people to understand them; yet we want to cry out whenever others try to categorize us. "I wanna be me. I'm like no one else that ever lived! Treat me for what I am. Do not assume that I am like any category of people!"

Science categorizes nature and makes generalizations about objects in nature. Social science does the same for human beings. Is this good for understanding? Does it contribute to stereotyping and inhumane treatment of those unlike ourselves? Here is the issue we are looking at here. The eleventh question is:

Should we generalize about people?

Categories and Generalizations

The Importance of Categories and Generalizations to Human Beings

Sociology is a social science, and therefore it makes generalizations about people and their social life. "The top positions in the economic and political structures are far more likely to be filled by men than by women." "The wealthier the individual, the more likely he or she will vote Republican." "In the United States the likelihood of living in poverty is greater among the African-American population than among whites." "American society is segregated." "Like other industrial societies, American society has a class system in which more than three-fourths of the population end up in approximately the same social class as what they were at birth."

But such generalizations often give me a lot of trouble. I know that the sociologist must learn about people and generalize about them, but I ask myself: "Are such generalizations worthwhile? Shouldn't we simply study and treat people as individuals?" An English professor at my university was noted for explaining to his class that "you should not generalize about people—that's the same as stereotyping and everyone knows that educated people are not supposed to stereotype. Everyone is an individual." (Ironically, this is *itself* a generalization about people.)

However, the more I examine the situation, the more I realize that all human beings categorize and generalize. They do it every day in almost every situation they enter, and they almost always do it when it comes to other people. In fact, we have no choice in the matter. "Glass breaks and can be dangerous." We have learned what "glass" is, what "danger" means, and what "breaking" is. These are all categories we apply to situations when we enter so that we can understand how to act. We generalize from our past. "Human beings who have a cold are contagious, and, unless we want to catch a cold, we should not get close to them." We are here generalizing about "those with colds," "how people catch colds," and "how we should act around those with colds." In fact, every noun and verb we use is a generalization that acts as a guide for us. The reality is that we are unable to escape generalizing about our environment. That is one aspect of our essence as human beings. This is what language does to us. Sometimes our generalizations are fairly accurate; sometimes they are unfounded. However, we do in fact generalize: all of us, almost all the time! The question that introduces this chapter is a foolish one. *Should we generalize about people* is not a useful question simply because we have no choice. A much better question is:

How can we develop accurate generalizations about people?

The whole purpose of social science is to achieve accurate categorizations and generalizations about human beings. Indeed, the purpose of almost all academic pursuits involves learning, understanding, and developing accurate categories and generalizations.

For a moment let us consider other animals. Most are prepared by instinct or simple conditioning to respond in a certain way to a certain stimulus in their environment. So, for example, when a minnow swims in the presence of a hungry fish, then that particular minnow is immediately responded to and eaten. The fish is able to distinguish that type of stimulus from other stimuli, and so whenever something identical to it or close to it appears, the fish responds. The minnow is a concrete object that can be immediately sensed (seen, smelled, heard, touched), so within a certain range the fish is able to easily include objects that look like minnows and to exclude those that do not. Of course, occasionally a lure with a hook is purposely used to fool the fish, and a slight mistake in perception ends the fish's life.

Human beings are different from the fish and other animals because we have *words for objects and events* in the environment, and this allows us to *understand* that environment and not just respond to it. With words we are able to make many more distinctions, and we are able to apply knowledge from one situation to the next far more easily. We are far less dependent on immediate physical stimuli. So, for example, we come to learn what fish, turtles, and whales are, as well as what minnows, worms, lures, and boats are. We read and learn what qualities all fish have, how fish differ from whales, and what differences fish have from one another. We learn how to catch fish, and we are able to apply what we learn to some fish but not other fish. We begin to understand the actions of all fish—walleyes, big walleyes, big female walleyes. Some of us decide to study pain, and we try to determine if all fish feel pain, if some do, or if all do not. Humans do not then simply respond to the environment, but they label that environment, study and understand that environment, develop categories and subcategories for objects in that environment, and constantly try to generalize from what they learn in specific situations about those categories. Through understanding a category we are able to see important and subtle similarities and distinctions that are not available to animals who do not categorize and generalize with words.

Generalizing allows us to walk into situations and apply knowledge learned elsewhere to understanding objects there. When we enter a classroom we know what a teacher is, and we label the person at the front of the room as a teacher. We know from past experience that teachers give grades, usually know more than we do about things we are about to learn in that classroom, have more formal education than we do, and usually resort to testing us to see if we learned something they regard as important. We might have also learned that teachers are usually kind (or mean), sensitive (or not sensitive), authoritarian (or democratic); or we might have had so many diverse experiences with teachers that whether a specific teacher is any of these things will depend on that specific individual. If we do finally decide that a given teacher is in fact authoritarian, then we will now see an "authoritarian teacher," and we will now apply what we know about such teachers from our past.

This is a remarkable ability. We are able to figure out how to act in situations we enter because we understand many of the objects we encounter there by applying relevant knowledge about them that we learned in the past. This allows us to intelligently act in a wide diversity of situations, some of which are not even close to what we have already experienced. If we are open-minded and reflective, we can even evaluate how good or how poor our generalizations are, and we can alter what we know as we move from situation to situation.

The problem for almost all of us, however, is that many of our generalizations are not carefully arrived at or accurate, and it is sometimes difficult for us to recognize this and change them. Too often our generalizations actually stand in the way of our understanding, especially when we generalize about human beings.

To better understand what human beings do and how that sometimes gets us into trouble, let us look more closely at what "categories" and "generalizations" are.

The Meaning of Categorization

Human beings categorize their environment; that is, we isolate a chunk out of our environment, distinguish that chunk from all other parts of the environment, give it a name, and associate certain ideas with it. Our chunks—or categories—arise in interaction; they are socially created. We discuss our environment, and we categorize it with the words we take on in our social life: "living things," "animals," "reptiles," "snakes," "poisonous snakes," "rattlers." A category is created, and once we understand it, we are able to compare objects in situations we encounter to that category. The number of distinctions we are able to make in our environment increases manyfold. It is not only nouns that represent categories (*men, boys*) but also verbs (*run, walk, fall*), adverbs (*slow, fast*), and adjectives (*weak, strong, intelligent, married*). Much of our learning is simply aimed at understanding what various categories mean, and this involves understanding the qualities that make up those categories and the ideas associated with them.

Through learning about people (a category) we come to recognize that "all people" possess certain qualities, some of which they share with other animals (cells, brains, reproductive organs), and some that seem to be unique to them (language, stereoscopic vision, conscience). We understand that people can be divided into young and old, white and black, men and women, single and married. Most of us have a pretty good idea of what a male is and a female. If asked, we could explain who belongs to the categories of homosexual and heterosexual. We do not simply recognize objects that do or do not belong; we *understand* the category by being able to describe the qualities we believe belong to objects that fit and objects that do not. We might say that a male has a penis, an old person is anyone older than 60, a teacher is someone who transmits knowledge, a human being is an animal who has a soul.

We argue over these definitions, and the more we understand, the more complex these definitions become. But categories and definitions are a necessary part of all of our lives. Armed with these, we go out and are able to cut up our environment in complex and sophisticated ways. We see an object and determine what it is (that is, what category it belongs in), and because we know something about that category, we are able to apply what we already know to that object. This allows us to act appropriately in many different situations. Simply think of all the people we meet in a given day, most of whom we know nothing about except for whatever we gather from a quick glance. We may note age, gender, dress, hairstyle, demeanor, or just a smile, and we quickly determine how to act. We are forced to place individuals into categories so we know what to do in a multitude of social situations.

It is necessary for all human beings to categorize, define, and understand their environment. (This statement is itself a generalization about all human beings.) If we are honest with ourselves, we should recognize that each of us has created or learned thousands—even tens of thousands—of categories that we use as we look at what happens around us. The purpose of a biology class is to create useful categories of living things so that we can better understand what these things are—how they are similar, how they differ from nonliving things, and how they differ from one another. Musicians, artists, baseball players, political leaders, students, parents, scientists, con artists, and police—all of us live our lives assuming certain things about our environment based on the categories we have learned in interaction with others.

A *role* is a category we use to understand the situations we encounter. A role is a set of expectations that people have of an actor in a position within a social situation. If you are a ticket taker, I expect you to ask me for my ticket; if you are an employee in the theater working behind a candy counter, I expect you to ask me if you can help me; if you are someone sitting in a theater watching a movie, I expect you to keep quiet. If you are a receptionist in a doctor's office, I expect you to tell me when I may see the doctor; if you are a nurse, I expect you to ask me a series of questions; and if you are a physician, I trust you to respect my body. Every set of role expectations I have for others and for myself is an attempt to categorize people. It helps them know what to do; it helps me know what they will do, and what I should do in relation to them. Such expectations are an inevitable part of our lives.

The Meaning of Generalization

A *category* is an isolated part of our environment that we notice. We generalize about that category by observing specific instances of objects included in it and by isolating common qualities that seem to characterize those included in that category, including other yet unobserved members we might observe in the future. We watch birds build a nest, and we assume that all birds build nests out of sticks (including birds other than the robins and sparrows we observed). We continue to observe and note instances where birds use materials other than sticks, and then we learn that some birds do not build nests but dig them out. More often, our generalizations are a mixture of observation and learning from others: we learn that wealthy people often drive Mercedes, and that police officers usually carry guns. On the basis of generalizing about a category, we are able to predict future events where that category comes into play. When we see a wealthy person, we expect to see a Mercedes (or something that we learn is comparable); and when we see a police officer, we expect to see a gun. That is what a generalization is.

A *generalization describes the category.* It is a statement that characterizes objects within the category and defines similarities and differences with other categories. "This is what an educated person is!" (in contrast to an uneducated person). "This is what wealthy people do to help ensure that privilege is passed down to their children." "This is what U.S. presidents have in common." "This is what Catholic people believe in."

As we shall see shortly, a generalization sometimes goes beyond just describing the category. It also explains why a particular quality develops. *That is, a generalization about a category will often be a statement of cause.* "Jewish people are liberal on social issues because of their minority position in Western societies." "U.S. presidents are male because ..." "Wealthy people send their children to private schools because ..."

Human beings, therefore, categorize their environment through words. On the basis of observation and learning, they come to develop ideas concerning what qualities are associated with those categories. They also develop ideas as to why those qualities develop. Ideas that describe the qualities that belong to a category and ideas that explain why those qualities exist is what we mean by *generalizations*.

The Stereotype

When it comes to people, generalization is very difficult to do well. The principal reason for this is that we are judgmental, and too often it is much easier for us to generalize for the purpose of evaluating (condemning or praising) others than for the purpose of understanding them. When we do this we fall into the practice of *stereotyping*.

A stereotype is a certain kind of categorization. It is a category and a set of generalizations characterized by the following qualities:

1. *A stereotype is judgmental.* It is not characterized by an attempt to understand, but by an attempt to condemn or praise the category. It makes a value judgment, and it has a strong emotional flavor. Instead of simple description of differences, there is a moral evaluation of those differences. People are judged good or bad because of the category.

2. *A stereotype tends to be an absolute category.* That is, there is a sharp distinction made between those inside and those outside the category. There is little recognition that the category is merely a guide to understanding and that, in reality, there will be many individuals within a category who are exceptions to any generalization.

3. *The stereotype tends to be a category that overshadows all others in the mind of the observer.* All other categories to

which the individual belongs tend to be ignored. A stereotype treats the human being as simple and unidimensional, belonging to only one category of consequence. In fact, we are all part of a large number of categories.

4. *A stereotype does not change with new evidence.* When one accepts a stereotype, the category and the ideas associated with it are rigidly accepted, and the individual who holds it is unwilling to alter it. The stereotype, once accepted, becomes a filter through which evidence is accepted or rejected.

5. *The stereotype is not created carefully in the first place.* It is either learned culturally and simply accepted by the individual or created through uncritical acceptance of a few concrete personal experiences.

6. *The stereotype does not encourage a search for understanding why human beings are different from one another.* Instead of seeking to understand the cause as to why a certain quality is more in evidence in a particular category of people, a stereotype aims at exaggerating and judging differences. There is often an underlying assumption that "this is the way these people are," it is part of their "essence," and there seems to be little reason to try to understand the cause of differences any further than this.

Stereotypes are highly oversimplified, exaggerated views of reality. They are especially attractive to people who are judgmental of others and who are quick to condemn people who are different from them. They have been used to justify ethnic discrimination, war, and systematic murder of whole categories of people. Far from arising out of careful and systematic analysis, stereotypes arise out of hearsay and culture, and instead of aiding our understanding of the human being, they always stand in the way of accurate understanding.

It is not always easy to distinguish a stereotype from an accurate category. It is probably best to consider stereotypes and their

opposites as extremes on a continuum. In actual fact, most categories will be neither perfectly accurate nor perfect examples of
stereotypes. There are, therefore, *degrees of stereotyping* that we should
recognize:

Stereotype	Accurate Categorization
Judgmental	Descriptive
No exceptions	Exceptions
All-powerful category	One of many
Rejects new evidence	Changes with evidence
Not carefully created	Carefully created
Not interested in cause	Interested in cause

One final point. Stereotypes, we have emphasized, are judgmental. They are meant to simplify people so that we know which
categories of people are good and which are to be avoided or condemned. This is the link between stereotypes and prejudice. Prejudice is an attitude toward a category of people that leads the actor to
discriminate against individuals who are placed in that category. Always the category is a stereotype (judgmental, absolute, central,
rigid, cultural, and uninterested in cause). When a prejudiced actor
identifies an individual in the category, a lot is assumed to be true
and disliked about that individual, and a negative response results.
And once he or she acts in a negative way toward that individual,
there is a ready-made justification: the stereotype ("I discriminate
because this is the way they are!"). Stereotypes are oversimplifica-
tions of reality, and they act as both necessary elements of prejudice
and rationalizations of it. Unfortunately, the stereotype also acts as a
set of role expectations for those in the category, and too often people who are judged negatively are influenced to judge themselves
accordingly.

Social Science: A Reaction to Stereotypes

Creating categories about people and generalizing intelligently is very difficult to do unless we work hard at it. A big part of a university education is to uncover and critically evaluate stereotypes in order to obtain a better understanding of reality. Each discipline in its own way attempts to teach the student to be more careful about categorizing and generalizing.

Because this book focuses on the perspective of sociology and social science, I would like to show how social science tries to rid us of stereotypes through the careful development of accurate categories and generalizations about human beings. Social science is a highly disciplined process of investigation whose purpose is to question many of our uncritically accepted stereotypes and generalizations. Social science does not always succeed. There are many instances of inaccuracies and even stereotyping that have resulted from poor science or from scientists simply not being sensitive to their own biases. It is important, however, to recognize that even though scientists make mistakes in their attempts to describe reality accurately, the whole thrust and spirit of social science is to control personal bias, to uncover unfounded assumptions about people, and to understand reality as objectively as possible. Here are some of the ways that social science (as it is supposed to work) aims at creating accurate categories and generalizations about human beings:

1. *Social science tries hard not to be judgmental about categories of people.* We recognize that generalizations and categories must not condemn or praise but must simply be guides to understanding. To stereotype is to emphasize qualitites in others that we dislike or to emphasize qualities in others that are similar to our own that we like. To say that a category of people is lazy is to stereotype; to say that a group has a higher unemployment rate than another is to carefully generalize. To say that a group is filthy rich or trashy is to stereotype; to say that a group has a higher average income than other groups is to carefully generalize. It is sometimes difficult to draw the line between a stereotype and a generalization about people, but, in general, the purpose of each is different: to simply generalize is to try to understand other people; to stereotype is to put

understanding aside in order to take a stand about other people, usually a negative one. Stereotyping prevents the individual from understanding reality.

I am not claiming that people should stay away from making value judgments about categories of people. We all have values we believe in, and we must include these when we act around others. I try to stay away from violent individuals; I try to change racists and sexists. I make judgments about students who plagiarize and employers who do not treat employees with respect. However, such judgments should be made carefully and explicitly (on the table), and only after categories and generalizations have been developed out of a process whose purpose is to *understand*. Good social science tries to separate making value judgments about people from understanding people, because if we do both simultaneously, stereotyping will inevitably be the result. It may be all right to condemn certain qualities in others, but it should be based on objective categorization, not stereotyping. Perhaps one goal for students should be to work toward developing informed value judgments.

2. *Categories and generalizations in social science are rarely—if ever—absolute.* Social scientists begin with the assumption that it is difficult to generalize about people and that every time we do exceptions are likely, and often a large number. By definition, all atheists do not believe in God, but there is absolutely nothing else we can say about all atheists. However, we might contend that atheists tend to be more educated (but there are many exceptions to this), male rather than female (but there are many exceptions to this), and raised by atheist parents (but there are many exceptions to this). We can tease out generalizations about atheists from carefully studying them, but we will never find a quality that all of them have other than their belief that there is no God. This goes for every category of people we try to understand: those who commit suicide, those who abuse drugs, those who commit violent acts against children, serial killers, and students who do not finish college. We can generalize, but we must be careful, and we must assume exceptions within every category we create. The scientific generalization is treated as a probability rather than an absolute. So we might say that among young

adults, less than 10 percent regularly use illegal drugs. Stereotyping, on the other hand, admits to few exceptions and involves rigid and absolute divisions between people, assuming that each individual within the category has the quality identified. ("Young people today are a bunch of drug addicts.") To claim that "Jewish people are rich" is to stereotype. To declare that Jewish people in the United States have the second highest per capita average income of any religious group is to generalize carefully. When averages are used to compare people, we must be probabilistic: we will recognize that there is a whole range of incomes among Jewish people, but that their average will simply be higher than most, lower than some. Whether a certain individual who is Jewish will be found to be wealthy is not easily predictable: most Jewish people are not wealthy, and many are poor.

Good social science even attempts to identify exactly how many exceptions within a category actually exist. We try to make the exceptions precise: In 1985, for example, 11.4 percent of the white population was poor (88.6 percent was not), as were 31.3 percent of the African-American population and 29 percent of the Spanish-origin population. We are unwilling to say that "everyone is getting divorced nowadays." Instead, we might say that "if the divorce rate remains as high as it is now, in the current generation of those who marry approximately one in two marriages will end in divorce" (there is, approximately, a 50 percent exception rate to "everyone getting divorced").

3. *Categories in social science are not assumed to be all-important for understanding the individual.* A stereotype is itself an assumption that a certain category necessarily dominates an individual's life. We might meet a young African-American single male artist. The role of each of these categories may or may not be important to the individual. For some individuals, being male or single or an artist will be most influential; for others it will be being African American. For those of us who stereotype by race, it will almost always be African American.

For social scientists, the human being is thought to exist within many categories, sometimes interrelated, each having a different level of importance depending on the individual. This makes placing individual people into easy categories much more difficult than placing other objects, including other animals. The contemporary debate over "gay lifestyle" highlights this issue. Whereas those who are interested in making value judgments about others emphasize the importance of the "gay" category in one's life, those who attempt to understand people recognize that this category is important for some and unimportant for others. People who are gay live in the world of the working or middle class too; the business, professional, or artistic worlds; the city or the rural community; the religious or the nonreligious community. Human beings should never be pigeonholed into one category if we are to be accurate.

4. *Social science tries to create categories and generalizations through carefully gathered evidence.* Stereotypes tend to be cultural; that is, they are taught by people around us who have generalized based on what they have simply accepted from others or what they have learned through personal experience (which is usually extremely limited in scope, unsystematic, subject to personal and social biases, and uncritically observed). Science tries hard to encourage accurate generalizations through making explicit how generalizations must be arrived at. In fact, those who stereotype rarely know exactly where their categorization has come from, and normally when they are pushed, they will admit that it is something they picked up or it is based on limited experience. Scientists, on the other hand, are supposed to know exactly where their generalizations have come from. They normally can point to evidence that has been derived from studies that have been reported and analyzed over and over again. Scientists—as well as most intellectuals—put their faith in process (how ideas are arrived at); most of the rest of us (who too often stereotype) rarely question the process by which we have arrived at our generalization; instead, we simply accept it.

5. *Generalizations in social science are tentative and subject to change* because evidence is constantly being examined. Stereotypes, on the other hand, are unconditionally held. Once held, a stereotype causes the individual to select out evidence that only reaffirms that stereotype. A stereotype resists change. When we believe that whites have superior abilities to nonwhites, we tend to notice only those indi-

viduals who support our stereotype. If we believe that politicians are selfish bureaucrats, we tend to forget all those political leaders who are unselfish and who get things done. (Note: the category "politician" gives away that one is stereotyping rather than simply generalizing, because politician has come to mean someone who is not worth our respect.) Because the purpose of a stereotype is to condemn or praise a category of people, it becomes difficult to evaluate evidence. The stereotype is embedded in the mind of the observer, it takes on an emotional flavor, and evidence that might contradict it is almost impossible to accept.

A generalization in social science about a category of people is subject to change as soon as new evidence is discovered. The final truth about people is never assumed to have been found. The generalization is always taken as a tentative guide to understanding rather than a quality that is etched in stone.

Our culture has taught us that "those who pray together stay together." This is a stereotype. We normally believe this if we are ourselves among those who pray. There is a strong urge among most of us to resist changing our generalization about those people of whom we are a part. Careful examination of the generalization may find that praying together may not have any effect on holding families together (or we might find that it has an effect on some), or we most probably will find that doing anything together as a family will have an effect on family stability. To generalize without stereotyping, however, involves holding onto generalizations only as tentative guides to reality that must be changed whenever there is new evidence. Those of us who stereotype take our categories far too seriously: we trust our categories far more than they deserve.

6. *Scientists do not categorize as an end in itself.* Instead, scientists categorize because they seek a certain kind of generalization: they seek to understand cause. In social science that means we seek to know *why* a category of people tends to have a certain quality. We generalize about categories of people *to better understand what causes* the existence of qualities that belong to a given category. We seek to understand the cause of schizophrenia, but we can only do this after we understand what characterizes those people who are schizophrenic. How is this category of people different from others? We

seek to know why poverty exists in this society. First, we must understand what poverty is (that is, we try to describe the lives of the people whom we label poor). What do these people have in common, if anything? How are their lives different from other people's? Then, what has led them to poverty? For example, how many came from families who were also poor? How many are single parents? How many are children? How many are products of corporate restructuring? How many have job skills that are no longer needed? Can we, through our studies, determine some of the social conditions that led to poverty among most or even a significant minority of these people? That is what social science tries to understand.

Those who stereotype are not normally interested in cause. The category stands by itself as important. Often, according to Roger Brown (1965:181–189), it is good enough to simply believe a certain quality is "part of their essence" and to ignore its cause. On the other hand, if a large number of Hispanic Americans do not finish high school, for example, then the social scientist is driven to understand why.

Real generalization in science therefore is to uncover why certain qualities make up a category, and why they are less in evidence in other categories. *Why* is there increasing individualism among Americans? *Why* do some people graduate college and not others? *Why* are women absent from the top political and economic positions in American life? *Why* is there an increase in the number of people who are experiencing downward social mobility in the United States? *Why* is there a rising suicide rate among young people? In every one of these cases we find a category, we describe those who make up that category, and we attempt to generalize as to why a certain quality exists in that category. To judge? No. True of everyone in the category? No. The only category of importance? No. A fixed category that clearly and absolutely distinguishes between one group and another? No. A generalization that we can regard as true without reservation? No.

Summary and Conclusion

Social science is sometimes misinterpreted by the public. Remember that those of us who stereotype seek evidence to support our stereotypes. We lay in the weeds—so to speak—waiting to pounce on any evidence that supports us (ignoring all evidence that does not). As careful as social science might be, what scientists find can be stretched a long way and misused. There is evidence, for example, that African Americans do less well on standardized intelligence tests than whites. For the social scientist this is a tentative generalization, it is puzzling, and it needs more understanding. The social scientist wants to know why and will look at any inherent bias in the test as well as the social conditions that might lead to this discrepancy. There is no sweeping, absolute generalization here. After all, we are talking about averages. There is no attempt to condemn or defend ethnic groups, to justify or rationalize racism. To the racist, however, this might become more evidence that whites are superior people. It might be used to reinforce a stereotype. This is why, unfortunately, people who try hard to understand categories of people carefully and objectively (as social scientists are supposed to do) become frustrated by those who exaggerate and twist what is found to fit their stereotypes.

Before we forget where we began this discussion, let me remind you once more:

☐ Human beings generalize.

☐ We must generalize.

☐ It is important to generalize carefully, and, when it comes to people, we should try to keep away from stereotyping if we want to understand them.

☐ Our generalizations about people must attempt to understand; our generalizations must be considered only tendencies among certain people; they must be accepted as open, tentative generalizations; and we must become aware of how we have arrived at our generalizations, always keeping in mind the importance of good evidence. Generalizations must also respect the complexity of the individual, and we should seek to understand why people differ and be suspicious of those who simply categorize in order to condemn.

Finally, we should examine once more the question that we started with: Should we generalize about people? This is not an easy question to answer.

We must begin our answer by admitting that we all take whatever we know about categories of people and apply it to situations we encounter. When we see that the individual is a child or an elderly person, male or female, single or married, a professor or a physician, wealthy or poor, kind or insensitive, that information guides us in our actions. If we are careful, we will recognize that our view of the other must be tentative, that the individual may in fact be an exception in our category, and that we must be ready to change whatever we think as we get to know that person as a unique individual. In fact the category we use may end up being unimportant for understanding this particular person.

In a society where we cry out for individual recognition, few of us will admit we want others to place us into categories and generalize. "Do not categorize me. I'm an individual!" Yet, if we are honest, we will recognize that those who do not know us will be forced to categorize us. It actually is not too bad if the category is a positive one. If we apply for a job we want the employer to categorize us as dependable, hardworking, knowledgeable, intelligent, and so on. We will even try to control how we present ourselves in situations so we can influence the other to place us in favorable categories: I'm cool, intelligent, sensitive, athletically talented, educated. The doctor may try to let people know "I am a physician" so that they will think highly of him or her as an individual. The individual who announces himself as a boxer is telling us that he is tough, the rock musician is telling us that she is talented, the minister that he or she is caring—in many such cases it does not seem so bad if we are being categorized. For almost all of us, however, it is the *negative* categorization that we wish to avoid. And this makes good sense: no one wants to be put into a category and negatively judged without having a chance to prove himself or herself as an individual.

But no matter how we might feel about others categorizing us and applying what they know to understanding us as a member of that category, the fact is that, except for those we know well, human beings can only be understood if we categorize and generalize. If we do this carefully, we can understand much about them, but if we are sloppy, we sacrifice understanding and end up making irrationally-based value judgments about people before we have an opportunity to know them as individuals.

We should not throw careful generalizations out the window in the name of treating all people as individuals. As much as every individual might deserve being treated as an individual, knowledge about anything—including human beings—is possible only through generalization. The HIV virus is spread through the transmission of bodily fluids through sexual contact, blood transfusion, or drug use—that is a generalization that can cause death if ignored. The history of African Americans has been one of active and subtle discrimination by the white community—that is a key to understanding many of the problems that are important in American society today. The upper class in American society has more privilege than any other class in the political, educational, and legal systems—that is an important generalization that sensitizes the individual to the limitations of our democracy. None of these generalizations is absolute, unbending, or certain, and none is meant to condemn or defend any category of people. They are not stereotypes.

Social science—and sociology as a social science—is an attempt to categorize and generalize about human beings and society, but always in a careful manner. Its purpose is to reject stereotyping. It is a recognition that generalizing about people is necessary and inevitable, but stereotyping is not.

If we have to generalize, let's try to be careful. Stereotyping does not serve our own interests well because it blocks understanding; nor does it help those we stereotype.

REFERENCES

Of course, almost any book in social science will be an example of categorization and generalization. It is difficult to find a source that specifically discusses the issues described in this chapter. Here are some works that have especially influenced me in understanding social science, stereotyping, categorizing, and generalizing.

Adorno, Theodor W., Else Frenkel-Brunswick, D. J. Levinson, and R. N. Sanford 1950 *The Authoritarian Personality*. New York: Harper and Row.

Allport, Gordon 1980 *The Nature of Prejudice*. Reading, MA: Addison-Wesley.

Aronson, Elliot 1988 *The Social Animal*. 5th ed. San Francisco: W. H. Freeman.

Babbie, Earl M. 1995 *The Practice of Social Research*. 7th ed. Belmont, CA: Wadsworth.

Becker, Howard S. 1973 *Outsiders*. Enlarged ed. New York: Free Press.

Berger, Peter L., and Thomas Luckmann 1966 *The Social Construction of Reality*. New York: Doubleday.

Blumer, Herbert 1969 *Symbolic Interactionism: Perspective and Method*. Englewood Cliffs, NJ: Prentice-Hall.

Brown, Roger 1965 *Social Psychology*. New York: Free Press.

Cohen, Morris R., and Ernest Nagel 1934 *An Introduction to Logic and Scientific Method*. New York: Harcourt Brace Jovanovich.

Ehrlich, Howard J. 1973 *The Social Psychology of Prejudice*. New York: Wiley Interscience.

Goffman, Erving 1961 *The Presentation of Self in Everyday Life*. New York: Doubleday (Anchor).

Goffman, Erving 1963 *Stigma: Notes on the Management of Spoiled Identity*. Englewood Cliffs, NJ: Prentice-Hall.

Hamilton, Davis L. 1981 *Cognitive Processes in Stereotyping and Intergroup Behavior*. Hillsdale, NJ: Lawrence Erlbaum Associates.

Hertzler, Joyce O. 1965 *A Sociology of Language*. New York: Random House.

Jacobs, Nancy R., Mark A. Segal, and Carol D. Foster 1988 *Into the Third Century: A Social Profile of America*. Wylie, TX: Information Aids.

Mills, C. Wright 1959 *The Sociological Imagination*. New York: Oxford University Press.

Ryan, William 1976 *Blaming the Victim*. Rev. ed. New York: Vintage.

Shibutani, Tamotsu 1970 "On the Personification of Adversaries." In *Human Nature and Collective Behavior*. Ed. Tamotsu Shibutani. Englewood Cliffs, NJ: Prentice-Hall.

Simpson, George E., and J. Milton Yinger 1985 *Racial and Cultural Minorities: An Analysis of Prejudice and Discrimination*. 5th ed. New York: Harper and Row.

Whorf, Benjamin Lee 1956 *Language, Thought, and Reality*. New York: John Wiley.

Index